ALGORITHMS IN ACTION

ALGORITHMS
IN ACTION

First Edition

Victor Savvich
Univeristy of Southern California

 cognella®
SAN DIEGO

Bassim Hamadeh, CEO and Publisher

Mieka Portier, Field Acquisitions Editor

Tony Paese, Project Editor

Alia Bales, Production Editor

Jess Estrella, Senior Graphic Designer

Trey Soto, Licensing Coordinator

Natalie Piccotti, Director of Marketing

Kassie Graves, Vice President of Editorial

Jamie Giganti, Director of Academic Publishing

Cover image copyright © 2018 iStockphoto LP/monsitj.

Printed in the United States of America.

3970 Sorrento Valley Blvd., Ste. 500, San Diego, CA 9212

Contents

Preface

T HIS BOOK IS AN INTRODUCTION TO algorithm design, intended to teach a variety of techniques for designing algorithms, with a focus on an intuitive understanding of algorithms. The book will train you on many basic algorithms, so you should be able to employ them in future algorithmic courses (like machine learning) and eventually apply them in your professional work. The book does not intend to be comprehensive nor complete.

The genesis of this book came about through the lecture notes I developed while teaching undergraduate and graduate computer science courses at Carnegie Mellon University and the University of Southern California. The book covers roughly a semester's worth of coursework, though some chapters go far beyond the standard lecture material, letting students dive deeper into the concepts and thus providing material to stimulate further thought and discussion. The book is intended for advanced undergraduate or master-level students in computer science and/or related technical disciplines. A foundation of undergraduate coursework in discrete mathematics, data structures, and calculus is highly recommended as a prerequisite. The book does not emphasize nor require programming, just pseudocode to encourage readers on conceptual understanding.

Analysis of algorithms is challenging for the most students, as they have not yet developed an experience in algorithmic problem solving. Students often easily come up with an erroneous intuitive solution, demonstrating their overconfidence in understanding material. My approach is to explain how to design algorithms, focusing on providing fundamental concepts, with a detailed, visual step-by-step algorithm execution. The book contains over 160 figures that help the reader to visualize the process. While I provide proofs of algorithm correctness, my goal is not to overwhelm the reader with rigorous mathematical proofs.

My hope with this book is to offer a reader-friendly approach to algorithms, with the numerous review questions and exercises at the end of each chapter (122 short review questions and 129 exercises in total) allowing readers to practice and apply the concepts taught.

Victor Savvich
Playa Vista, California
April 2019

Chapter 1

Review

I N THIS CHAPTER WE REVIEW BASIC concepts, from asymptotic complexity, graph theory, and mathematical proof techniques, as they are required for better understanding for the chapters that follow. If the reader has some previous acquaintance with these topics, the chapter should be enough to get started. If the reader has no previous background in these, we suggest a more thorough introduction such as *Mathematics for Computer Science* by Eric Lehman, Thomson Leighton and Albert Meyer.[1]

1.1 Runtime Complexity

The term *analysis of algorithms* is used to describe approaches to study the performance of algorithms. With each algorithm we associate a sequence of steps comprising this algorithm. We measure the run time of an algorithm by counting the number of steps and therefore define an algorithmic complexity as a numerical function $T(n)$, where n is the input size. Consider a problem of addition of two n-bit binary numbers. Let $T(n)$ represent an amount of time addition used to add two n-bit numbers. We want to define "time" $T(n)$ taken by the method of addition without having to worry about implementation details. The process of addition consists of the following two steps:

- Adding 3 bits (one bit is a carry bit)
- Writing down 2 bits (again, one bit is a carry bit)

---

1   Eric Lehman, Thomson Leighton, and Albert Meyer, *Mathematics for Computer Science,* ([Great Britain: Samurai Media Limited, 2017).

On any computer, adding and writing two bits can be done in constant time. By constant time we mean that the time is independent of the input size. Therefore, the total time of addition of two n-bit binary numbers is $T(n) = n \cdot c$, where the constant c can be different on different computers. We say that bit addition is a linear time algorithm. The process of abstracting away details and determining the rate of resource usage in terms of the input size is one of the fundamental ideas in computer science. In this course we will perform the following types of analysis:

1. The worst-case complexity (the maximum number of steps taken on any input)
2. The best-case complexity (the minimum number of steps taken on any input)
3. The average case complexity (the average number of steps taken on a random input)
4. The amortized time complexity (the average complexity over a sequence of operations)

We measure the runtime of an algorithm using following asymptotic notations: O, Ω, Θ.

1.1.1 Upper Bound (Big-O)

For any monotonic functions, f, g from the positive integers to the positive integers, we say $f(n) = O(g(n))$ (or $f(n) \in O(g(n))$) if $g(n)$ eventually dominates $f(n)$. Figure 1.1 helps you to visualize this relationship.

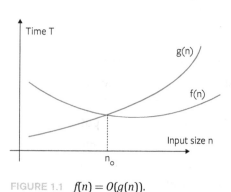

Formally, there exists a positive real number, $c > 0$, and a real number, n_0, such that $f(n) \leq c \cdot g(n)$ for all $n \geq n_0$.

Example: $n^2 + 2n + 1 = O(n^2)$. Since $n^2 + 2n + 1 \leq n^2 + 2n^2 + n^2 = 4n^2$ for $n \geq 1$, we choose $c = 4$ and $n_0 = 1$.

1.1.2 Lower Bound (Big-Omega)

For any monotonic functions, f, g from the positive integers to the positive integers, we say $f(n) = \Omega(g(n))$ (or $f(n) \in \Omega(g(n))$) if $f(n)$ eventually dominates $g(n)$. Formally, there exists a positive real number, $c > 0$, and a real number, n_0, such that $f(n) \geq c \cdot g(n)$ for all $n \geq n_0$.

FIGURE 1.1 $f(n) = O(g(n))$.

Example: $n^2 + 2n + 1 = \Omega(n^2)$. Since $n^2 + 2n + 1 \geq n^2$ for $n \geq 1$, we choose $c = 1$ and $n_0 = 1$.

1.1.3   Exact Bound (Big-Theta)

For any monotonic functions, f, g from the positive integers to the positive integers, we say $f(n) = \Theta(g(n))$ (or $f(n) \in \Theta(g(n))$) if $f(n) = O(g(n))$ and $f(n) = \Omega(g(n))$. Formally, there exists positive real numbers, c_1 and c_2, and a real number, n_0, such that $c_1 \cdot g(n) \leq f(n) \leq c_2 \cdot g(n)$ for all $n \geq n_0$.

Example: $n^2 + 2n + 1 = \Theta(n^2)$.

1.2   Lower Bound for Sorting

We will show here that any deterministic comparison-based sorting algorithm must take $\Omega(n \log n)$ time to sort an array of n elements in the worst case. Comparison-based sorting algorithms operate on the input by comparing pairs of elements. For example, Mergesort and insertion sort are comparison-based sorting algorithms. But bucket sort and radix sort are not. In order to show $\Omega(n \log n)$ bound we will play the Guess-a-Number game. The computer will select a number, x, between 1 and 10, and you need to determine x by asking questions. You'll keep guessing numbers until you find x. The guessing game can be viewed abstractly as a binary search tree (also called a decision tree). Figure 1.2 shows that we can guess any number between 1 and 10 by asking at most four questions.

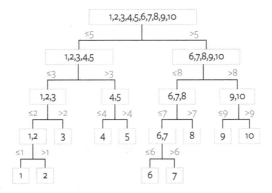

FIGURE 1.2   A binary search tree to guess a number between 1 and 10.

We will be using a decision tree to model the execution of any comparison-based sort. The execution of the sorting algorithm corresponds to tracing a path from the root of the decision tree to a leaf. At each internal node we make a comparison; based on that comparison we proceed further down to either the left or right subtree. Figure 1.3 depicts sorting an array of three elements [a, b, c]. In that tree each leaf represents a permutation of [a, b, c]. Generally, for sorting an array of n elements, each leaf in the

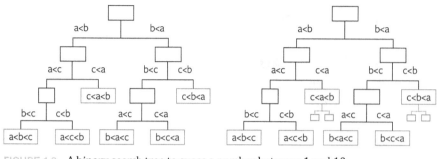

FIGURE 1.3   A binary search tree to guess a number between 1 and 10.

decision tree represents a permutation of the n elements. Hence, there are $n!$ leaves in the tree. Let us denote the tree by B. Next, we make the tree complete by adding extra nodes, as seen in figure 1.3. We will denote the new tree by B^* and its height by h. Since tree B^* is a complete binary tree, it has 2^h leaves. By construction, B^* has more leaves than B. It follows, $2^h \geq n!$, or, after taking the log of both parts, $h \geq \log(n!)$.

The height of the decision tree is the number of comparisons in a sorting algorithm; in other words, h is the runtime complexity $T(n)$ of sorting. Thus, $T(n) \geq \log(n!)$. Lastly, we simplify $\log(n!)$ as follows:

$$\log(n!) = \log(n(n-1)(n-2)\ldots 1)$$
$$\geq \log(n(n-1)\ldots(n-n/2))$$
$$\geq \log((n/2)^{n/2}) = \Omega(n \log n)$$

We have proved that any comparison-based sorting algorithm needs $\Omega(n \log n)$ comparisons. This holds even for a quantum computer!

1.3   Trees and Graphs

A graph G is a pair (V, E), where V is a set of vertices (or nodes) and E is a set of edges connecting the vertices. A self-loop is an edge that connects to the same vertex twice. A multi-edge is a set of two or more edges that have the same two vertices. A graph is simple if it has no multi-edges or self-loops. We always assume simple graphs unless otherwise noted. Graphs could be directed and undirected and weighted and unweighted (weights will usually be edge weights).

Theorem. *Let G be a graph with V vertices and E edges. The following statements are equivalent:*

1. *G is a tree.*
2. *Every two vertices of G are connected by a unique path.*
3. *G is connected and $V = E + 1$.*

4. *G is acyclic and $V = E + 1$.*
5. *G is acyclic and if any two non-adjacent vertices are joined by an edge, the resulting graph has exactly one cycle.*

To prove this, it suffices to show $1 \Rightarrow 2 \Rightarrow 3 \Rightarrow 4 \Rightarrow 5 \Rightarrow 1$. We'll just show $1 \Rightarrow 2 \Rightarrow 3 \Rightarrow 4$ and leave the rest to the reader.

Proof of $1 \Rightarrow 2$. We prove it by contradiction. Assume G is a tree that has two vertices connected by two different paths like in figure 1.4.

FIGURE 1.4 Two vertices connected by two different paths.

Then there exists a cycle! It follows that G cannot be a tree: a contradiction. ∎

Proof of $2 \Rightarrow 3$. Since every two vertices in G are connected by a path, G is a connected graph. We prove that in G the number of nodes and edges are related by $V = E + 1$. The proof is by strong induction on the number of nodes.

Base case: $V = 2$. Since a graph is simple, $E = 1$. Thus, $V = E + 1$.

Inductive hypothesis: Assume $V = E + 1$ for every graph with $V < n$ vertices.

Inductive step: Prove $V = E + 1$ for every graph with $V = n$ vertices.

Graph G has n vertices. We will use notation $V(G) = n$. We choose two adjacent vertices, x and y. We know that every two vertices in G are connected by a unique path. It follows that x and y are joined by an edge, like in figure 1.5.

Note in both subgraphs G_1 and G_2 the number of vertices is less than n. Indeed, G_1 does not contain vertex y, and G_2 does not contain vertex x. We can apply the inductive hypothesis to G_1 and G_2. It follows,

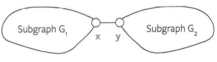

FIGURE 1.5 Graph G consists of two subgraphs G_1 and G_2.

$$V(G) = V(G_1) + V(G_2) = E(G_1) + 1 + E(G_2) + 1 = E(G) + 1.$$

This concludes the proof. ∎

Proof of $3 \Rightarrow 4$. We prove that G is an acyclic graph by contradiction. Assume that G has a cycle with k vertices in it. This cycle also contains k edges. Now let us count edges in the whole graph. We claim the number of edges in the graph will be at least V. Indeed, there are k edges in the cycle and at least $V - k + 1$ outside the cycle. ■

Theorem. *In an undirected simple graph $G = (V, E)$, there are at most $V(V-1)/2$ edges. In short, by using the asymptotic notation, $E = O(V^2)$.*

Proof. Choose any vertex in G. The possible number of edges leaving this vertex is $V - 1$. Take another vertex (different from the previous one). The possible number of edges leaving that vertex is $V - 2$ (don't count the edge between two vertices twice!), and so on. We have that the total number of edges is at most

$$(V - 1) + (V - 2) + \ldots + 2 + 1 = V(V-1)/2$$

This concludes the proof. ■

We define a *dense* graph $G = (V, E)$ as a graph in which the number of edges is $E = \Omega(V^2)$. We say that a graph is *sparse* if $E = O(V)$.

1.3.1 Graph Traversals

Graphs traversal means visiting all vertices in a systematic order. We can choose any vertex as a starting point. Then we will systematically enumerate all vertices accessible from it. Because a graph might contain cycles, we need some way for marking a vertex as having been visited. To do so we will keep a Boolean array, with all elements initially set to false. We will set a correspondent element to true as soon as we visit a particular vertex. Also, we need to keep in mind that the graph might be disconnected. There are two most common traversals:

- Depth-first search (DFS)

- Breadth-first search (BFS)

DFS uses a stack data structure for backtracking. BFS uses a FIFO queue for bookkeeping. Here is a pseudocode:

```
for all v in V do visited[v] = false
for all v in V do if !visited[v] traversal(v)
  traversal(v) {
    visited[v] = true
    for all w in adj(v)
        do if !visited[w] traversal(w)
  }
```

The runtime complexity of traversal is $O(V + E)$. There are two important properties of traversal: (1) It visits all the vertices in the connected component; (2) edges labeled by traversal form a spanning tree of the connected component.

1.3.2   Topological Sort

Suppose each vertex represents a task that must be completed and a directed edge (u, v) indicates that task v depends on task u. That is, u must be completed before v. If G is a direct acyclic graph (DAG), then there exists a valid order in which you can complete the tasks. This is called topological order or topological sort. If the graph is cyclic, no topological order exists. Consider the graph in figure 1.6.

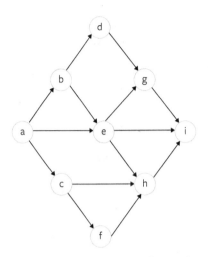

FIGURE 1.6   A directed acyclic graph.

The following sequence {a, b, c, d, e, f, g, h, i} is a valid topological sort. In other words, a topological order means arranging the vertices along a line so that all edges go from left to right. It should be evident from figure 1.6 that a topological sort is not unique. The following list {a, c, b, f, e, d, h, g, i} is another topological order.

The algorithm of finding a topological sort is based on traversal: run DFS (from any vertex) and return a vertex that has no undiscovered leaving edges. In figure 1.6, the possible DFS run may be $a{\rightarrow}e{\rightarrow}i$, making i the first vertex with no undiscovered leaving edges. From i, backtrack to e and then go to g. This makes g another vertex with no undiscovered leaving edges. From g, backtrack to e and then go to h. This makes h the third vertex with no undiscovered leaving edges. And so on. The algorithm will produce a topological order in reverse. Note, if we start DFS at any other vertex but a, we will need another run of DFS. The runtime complexity of the algorithm is linear $O(V + E)$.

1.3.3   Planar Graphs

A connected graph is planar if it can be drawn in the plane with each edge drawn as a continuous curve such that no two edges cross. There are many examples of planar graphs: any tree is planar, every cycle is planar, a complete graph K_4 is planar.

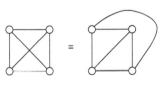

FIGURE 1.7   An example of a planar graph.

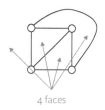

4 faces

FIGURE 1.8 A planar graph with four faces.

A planar graph in addition to vertices and edges also has disjoint faces.

Theorem. (Euler's formula) *If G is a connected planar graph with V vertices, E edges, and F faces, then*

$$V - E + F = 2.$$

Proof. The proof by induction on the number of edges.

Base case: $E = 1$. The identity holds, since $V = 2$ and $F = 1$.

Inductive hypothesis: Assume it's true for any graph with no more than E edges.

Inductive step: Prove it for graphs with E edges.

Start with a graph G that has E edges and remove one edge. There are two cases to consider:

1. The edge to remove lies on a cycle. See figure 1.9.
2. The edge to remove does not lie on a cycle. See figure 1.10.

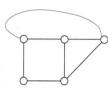

FIGURE 1.9 A graph in case 1.

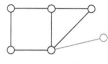

FIGURE 1.10 A graph in case 2.

In case (1) by removing an edge (in red) on a cycle, we obtain a new graph G' with $E - 1$ edges and F-1 faces. Since G' has $E - 1$ edges, the relation works by the induction hypothesis. That is $V - (E - 1) + (F - 1) = 2$. This simplifies to $V - E + F = 2$.

In case (2) by removing an edge (in red) that does not lies on a cycle, we obtain a new graph G' with $E - 1$ edges and $V - 1$ vertices. Since G' has $E - 1$ edges, the relation works by the induction hypothesis. That is $(V - 1) - (E - 1) + F = 2$. This simplifies to $V - E + F = 2$.

This completes the proof. ■

Theorem. *In any simple connected planar graph with at least 3 vertices, $E \leq 3V - 6$.*

Proof. If a graph has no cycles, then

$$E = V - 1 \leq V \leq V + (2V - 6) = 3V - 6,$$

since $V \geq 3$, and therefore $2V - 6 \geq 0$.

Assume a graph with cycles. We will count the number of pairs (edge, face) (i.e., Σ(edge, face)). Since each face is bounded by at least 3 edges, then Σ(edge, face) ≥ $3F$. Since each edge is associated with at most 2 faces, then Σ(edge, face) ≤ $2E$. Combining these two inequalities, we find $3F \le 2E$. But we know from the previous theorem that $V - E + F = 2$. It follows,

$$6 = 3V - 3E + 3F \le 3V - 3E + 2E = 3V - E.$$

Thus, we conclude $E \le 3V - 6$. ■

Corollary. *A simple connected planar graph with at least 3 vertices has a vertex of degree 5 or less.*

Proof. We know that in any graph, the sum of the degrees of all vertices is equal to twice the number of edges, $\Sigma \text{degree}(v) = 2E$. From the previous theorem, $2E \le 6V - 12$. Thus, the average degree is at most 6:

$$\frac{1}{V} \sum_{v \in V} \text{degree}(v) \le \frac{6\,V - 12}{V} = 6 - \frac{12}{V}$$

It follows there exists a vertex of degree 5 or less. ■

1.3.4 Coloring Planar Graphs

Given a planar graph, how many colors do you need in order to color the vertices so that no two adjacent vertices get the same color? Back in the 1880s, Francis Guthrie conjectured that four colors suffice. In 1976 K. Appel and W. Haken, using a special-purpose computer program, have proved that conjecture.

FIGURE 1.11   A graph coloring problem.

We won't prove the conjecture but let us prove the six-color theorem.

Theorem. (6-color theorem) *Every planar graph can be colored with at most six colors.*

Proof. By induction on the number of vertices.

Base case: If a graph has six or less vertices, then the result is obvious.

Inductive hypothesis: Assume that all graphs with $V - 1$ vertices are six-colorable.

Inductive step: Prove it for any graph with V vertices.

By previous corollary any graph G with V vertices has at least one vertex of degree 5 or less. Remove it from G. The remaining graph is planar and by induction can be colored with at most 6 colors. Now insert that vertex back. Since this vertex has at most 5 neighbors then at least one of 6 colors is not used. We color the vertex with one of the unused colors. ∎

1.3.5 Bipartite Graphs

A graph is bipartite if the vertices can be partitioned into two disjoint sets, X and Y, such that all edges go only between X and Y (no edges go from X to X or from Y to Y).

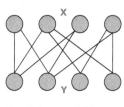

A complete bipartite graph (denoted by $K_{n,m}$, where n and m are sizes of two sets) is a special kind of *bipartite graph* where every vertex of the first partition is connected to every vertex of the second partition.

FIGURE 1.12   A bipartite graph.

Theorem. *A graph is bipartite if and only if it does not contain an odd length cycle.*

Proof. Note, the length of a cycle is the number of vertices in the cycle.

$\Rightarrow$) In a bipartite graph every cycle has vertices that must alternate between two partitions. Since the number of vertices in such a cycle is even, it is an even cycle.

$\Leftarrow$) Assume connected graph G has no odd cycle. Pick any vertex v. Define two sets of vertices based on parity of distance (even or odd) from v:

$$X = \{u \in V \mid d(v, u) \text{ is even}\}$$
$$Y = \{u \in V \mid d(v, u) \text{ is odd}\}$$

These sets provide a bipartition. If G had an odd cycle, then there will be a vertex present on both sets. Finish the proof for disconnected graph G. ∎

One of the famous problems on bipartite graphs is a matching problem. A subset of edges is a *matching* if no two edges have a common vertex. A *maximum matching* is a matching with the largest possible number of edges. Our goal is to find the maximum matching in a graph. We will show in chapter 7 that the problem of finding the maximum matching can be reduced to the maximum flow problem.

1.3.6   Other Famous Graph Problems

A Euler path is a path that uses each edge of a graph exactly once. A Euler cycle is a cycle that uses every edge of a graph exactly once. A graph that contains a Euler cycle is called a *Eulerian graph.*

Theorem. *A connected graph G is a Eulerian graph if and only if all vertices of G are of even degree.*

Proof. ⇒) Let $G = (V, E)$ be a Euler graph. Thus, G contains a Euler path. Let us walk that path. Visiting an intermediate vertex in the path contributes two to the degree of that vertex. It follows that every intermediate vertex has an even degree.

⇐) Assume that all vertices of G are of even degree. We construct a cycle starting at an arbitrary vertex v, going through the edges of G only once. Since every vertex is of even degree, we eventually come back to v. If this cycle includes all the edges of G, then G is a Eulerian graph. If not, we remove from G the cycle we have constructed. We will get a connected subgraph G_0 in which all vertices are of even degree. We again construct a new cycle in G_0. This process is repeated until we obtain a cycle that traces all the edges of G. We showed that G is a collection of cycles, hence G is a Eulerian graph. ∎

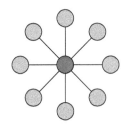

FIGURE 1.13   The *vertex cover* problem.

A *Hamiltonian path* is a path that visits each vertex of a graph exactly once. A Hamiltonian cycle is a cycle that visits every vertex in a graph exactly once (except for the start and end vertices). A graph that contains a Hamiltonian cycle is called a *Hamiltonian graph.* A problem closely related to Hamiltonian cycle is the famous *traveling salesman problem. Given a* weighted graph, find the shortest weighted Hamiltonian cycle. We will prove in chapter 9 that both problems are *unlikely to be solved in polynomial time.* The *traveling salesman problem* is one of the most intensively studied problems in computer science.

A *vertex cover* of an undirected graph is a subset of *vertices* such that for every edge (u, v) either u or v is in a *vertex cover.* The *vertex cover* problem is to find the minimum size *vertex cover.*

Given a graph, we say that a subset of vertices is independent if no two of them are joined by an edge. Given an undirected graph, the *independent set* problem is to find the largest independent set. Vertex cover and independent set are very closely related graph problems; see Exercise 15.

Two graphs, $G_1 = (V_1, E_1)$ and $G_2 = (V_2, E_2)$, are *isomorphic* if there is a bijective function $f: V_1 \rightarrow V_2$ such that an edge $(u, v) \in E_1$, if and only if an edge $(f(u), f(v)) \in E_2$. Two isomorphic graphs look differently but are structurally the same, up to the renaming of the vertices.

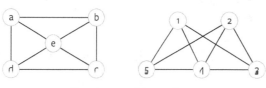

FIGURE 1.14   Two isomorphic graphs.

In figure 1.14 we can match vertices as follows: $a - 1, b - 2, c - 3, d - 5, e - 4$. Although we matched vertices in one particular way, there could be several ways to do it. Determining whether two graphs are isomorphic is not an easy task; however, computer scientists believe the problem can be solved in polynomial time.

REVIEW QUESTIONS

1. Mark the following assertions as TRUE or FALSE. No need to provide any justification.

    a. $n = O(n^2)$

    b. $n = O(\sqrt{n})$

    c. $\log n = \Omega(n)$

    d. $n^2 = \Omega(n \log n)$

    e. $n^2 \log n = \Theta(n^2)$

    f. $7 \log^2 n + 2n \log n = \Omega(\log n)$

    g. $5n \log n + 1024 n \log (\log n) = \Theta(n \log n)$

    h. $2^n + 100n^2 + n^{100} = O(n^{101})$

    i. $(1/3)^n + 100 = O(1)$

2. (T/F) Any function which is $\Omega (\log n)$ is also $\Omega (\log(\log n))$.

3. (T/F) If $f(n) = \Theta(g(n))$ then $g(n) = \Theta(f(n))$.

4. (**T/F**) If $f(n) = \Theta(g(n))$ then $f(n) = \Omega(g(n))$.

5. (**T/F**) If $f(n) = \Omega(g(n))$ then $2^{f(n)} = \Omega(2^{g(n)})$.

6. (**T/F**) BFS can be used to find the shortest path between any two nodes in a non-weighted graph.

7. (**T/F**) A DFS tree is never the same as a BFS tree.

8. (**T/F**) Algorithm A has a running time of $O(n^2)$ and algorithm B has a running time of $O(n \log n)$. From this we conclude that A can never run faster than B on the same input.

9. (**T/F**) Planar graph is a sparse graph.

10. (**T/F**) Every DAG contains a vertex with no incoming edges.

EXERCISES

1. Prove $g(n) = \Omega(f(n))$ if and only if $f(n) = O(g(n))$.

2. Prove or disprove $f(n) = O(g(n))$ implies $2^{f(n)} = O(2^{g(n)})$.

3. Arrange the following functions

$$\log n^n, \ n^2, \ n^{\log n}, \ n \log \log n, \ 2^{\log n}, \ \log^2 n, \ n^{\sqrt{2}}$$

in increasing order of growth rate, with $g(n)$ following $f(n)$ in your list if and only if $f(n) = O(g(n))$.

4. Arrange the following functions

$$4^{\log n}, \ \sqrt{\log n}, \ n^{\log \log n}, \ (\sqrt{2})^{\log n}, \ 2^{\sqrt{2 \log n}}, \ n^{1/\log n}, \ (\log n)!$$

in increasing order of growth rate with $g(n)$ following $f(n)$ in your list if and only if $f(n) = O(g(n))$.

5. What is the Big-O runtime complexity of the following function?

```
void bigOh1 (int n):
   for i=1 to n
      j=1;
      while j < n
         j = j*2;
```

6. What is the Big-O runtime complexity of the following function?

```
void bigOh2 (int n):
  if(n == 0) return "a";
  string str = bigOh2(n-1);
  return str + str;
```

7. What is the Big-Theta runtime complexity of the following function? Here find _ max finds the maximum element in the array L[0], L[1], ..., L[n − 1].

```
void bigTheta (int[] L, int n):
  while (n > 0)
    find _ max(L, n);
      n = n/4
```

8. The complete graph on n vertices, denoted K_n, is a simple graph in which there is an edge between every pair of distinct vertices. What is the height of the DFS tree for the complete graph K_n? What is the height of the BFS tree for the complete graph K_n?

9. We are interested in finding a simple path in a directed acyclic graph that visits all vertices once and only once. Design a linear time algorithm to determine if there is such a path in a given DAG.

10. Prove that a complete graph K_5 is not a planar graph.

11. Prove that a complete bipartite graph $K_{3,3}$ is not a planar graph.

12. In a connected bipartite graph, is the bipartition unique? Justify your answer.

13. Given a directed graph $G = (V, E)$ and a particular node $v \in V$, design a linear time algorithm to determine whether v is in a triangle of edges (a cycle of length 3).

14. Design a linear time algorithm which, given an undirected graph $G = (V, E)$ and a particular edge $e \in E$, determines whether G has a cycle containing e.

15. Given an undirected graph $G = (V, E)$, prove that S is an independent set if and only if $V − S$ is a vertex cover.

Chapter 2

Amortized Analysis

I N A SEQUENCE OF OPERATIONS, THE worst-case time does not occur often in each operation; some operations may be cheap, some may be expensive. Therefore, a traditional worst-case per operation analysis can give an overly pessimistic bound. When the same operation takes different times, how can we accurately calculate the runtime complexity? Amortized analysis gives the average performance (over time) of each operation in the worst case. Amortized analysis is not average case analysis. In average case analysis we compute the expected cost of each operation. Amortization is a technique used by accountants to average a large one-time expense over a long period of time. There are generally three methods for performing amortized analysis:

1. The aggregate method computes the upper bound $T(n)$ on the total cost of n operations. The amortized cost is given by $T(n)/n$. In this method each operation will get the same amortized cost, even if there are several types of operations in the sequence.
2. The accounting method (or the banker's method) computes the individual cost of each operation. We assign different charges to each operation; some operations may charge more or less than they actually cost. The amount we charge an operation is called its amortized cost.
3. The potential method (or the physicist's method). We won't use a potential method in this course.

2.1   Unbounded Array

The general implementation strategy: We maintain an array of a fixed length limit and an internal index size, which tracks how many elements are actually used in the array. When we add a new element, we increment size; when we remove an element, we decrement size. How do we proceed when the array is full and we need to add another element? At that point, we allocate a new array twice as large and copy the elements we already have to the new array. So, if the current array is full, the cost of insertion is linear; if it is not full, insertion takes a constant time. In order to make the analysis as concrete as possible, we will count the total number of inserts and the number of copy operations. In this model we won't analyze deletions (see exercise 3 for insertions and deletions). In table 2.1, we record the current size of the array, its new size, the number of insets, and the number of copies. The table shows that 9 inserts require $1 + 2 + 4 + 8 = 15$ copy operations. Therefore, the amortized cost of a single insert is the total cost $(9 + 15 = 24)$ over 9 inserts, which is 2.67.

TABLE 2.1 The cost of insertions

Insert	Old size	New size	Copy
1	1	—	—
2	1	2	1
3	2	4	2
4	4	—	—
5	4	8	4
6	8	—	—
7	8	—	—
8	8	—	—
9	8	16	8

Let us generalize the pattern. Assume we start with the array of size 1 and make $2^n + 1$ inserts. These inserts will require $1 + 2 + 4 + ... + 2^n = 2^{n+1} - 1$ copy operations. Thus, the total work (inserts plus copies) is given by $(2^n + 1) + (2^{n+1} - 1) = 3 \times 2^n$. Next, we compute the average cost per insert as a limit when the input size tends to infinity:

$$\lim_{n \to \infty} \frac{3 \cdot 2^n}{1 + 2^n} = 3$$

We say that the amortized cost of insertion is constant, namely $O(3)$. Such method of analysis is called an *aggregate* method. The aggregate method seeks an upper bound on the total running time of a sequence of operations.

Let us compute the amortized cost of insertion using the *accounting* method. This method seeks a payment for each individual operation. Intuitively, we maintain a bank account and each operation is charged to it. Some operations are charged very little but also generate a surplus. Others drain the savings. The balance in the bank account must always remain positive.

We will assign a dollar token to each operation. It costs a token to insert an element and another token to move it when we need to double the array size. It follows we have to assign at least 2 tokens to each insert: we pay one token to perform an operation, and we put one token into the bank. Figure 2.1 demonstrates the insertion process starting with an array of size one and an empty bank account.

In that picture we see that after third insertion, the bank account is empty, and after fourth insertion, the bank account has only one token. On the next, fifth insertion, we need to double the array size from 4 to 8. Clearly, we do not have enough money in the bank to pay for it.

Let us increase the number of tokens for each insert to three tokens: we pay one token to perform an operation, and we put two tokens into the bank. Figure 2.2 demonstrates the insertion process.

Now the bank has enough money to perform fifth insertion. In the next few inserts (6th–8th) we generate surplus.

In the next insert, we drain our savings. How do we know there will be enough money in the bank to pay for moving when we need to double the array size?

Doubling the array size say from N to $2N$, we need at least N tokens in the bank. Those N extra tokens will be generated by $N/2$ new inserts. Therefore, assigning three tokens per insert, we were able to pay for all the operations. This proves that our amortized cost is at most three.

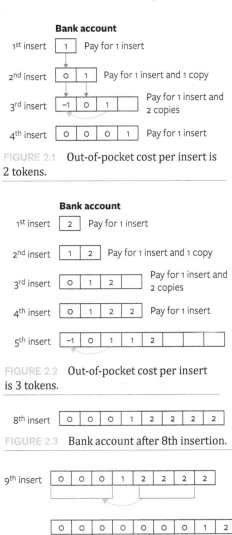

Bank account

FIGURE 2.1 Out-of-pocket cost per insert is 2 tokens.

FIGURE 2.2 Out-of-pocket cost per insert is 3 tokens.

FIGURE 2.3 Bank account after 8th insertion.

FIGURE 2.4 Bank account after 9th insertion.

2.2 Binary Counter

Given a binary number with $\log(n)$ bits, stored as an array, where each entry $A[i]$ stores the i-th bit, the cost of incrementing a binary number is the number of bits flipped. We use the standard way of incrementing the counter, which is to toggle the lowest order bit. What is the amortized cost per increment? As an example, consider 3-bit numbers and count the number of flips for each increment.

Table 2.2 shows that incrementing 000 requires a single flip, incrementing 001 results in two flips, and incrementing 111 results in 3 flips.

Clearly, in the worst-case all bits are flipped, so the cost per increment is $O(\log n)$. Now suppose we increment n times, starting with a zero-binary number. If we only use the worst-case running time for each increment, we get an upper bound of $O(n \log n)$. Although this bound is correct, we can do better.

TABLE 2.2 The number of flips for 3-bit numbers

	# of flips
000	1
001	2
010	1
011	3
100	1
101	2
110	1
111	3

2.2.1 The Aggregate Method

Let us think about how often we flip a single bit. Consider the least significant bit. Each time we increment a binary number, that bit is changed. Thus, the number of times the bit changes is n. Consider the next significant bit. How often is it toggled? $n/2$ times. The next bit is toggled $n/4$ times, and so on. The most significant bit is toggled only twice. Thus, the total cost is given by

$$n + \frac{n}{2} + \frac{n}{4} + \ldots + 2 = n\left(1 + \frac{1}{2} + \frac{1}{4} + \ldots + \frac{2}{n}\right) \leq n \sum_{k=0}^{\infty} \frac{1}{2^k} = 2n.$$

It follows the amortized cost per increment is $O(2)$.

2.2.2 The Accounting Method

The key point to observe is that each increment has exactly one $0 \rightarrow 1$ flip. But different increments have different numbers of $1 \rightarrow 0$ flips. Our accounting policy is the following: Every time you flip $0 \rightarrow 1$, pay the actual cost of 1, plus put 1 into a bank; every time you flip $1 \rightarrow 0$, use the money in the bank to pay for that flip. Consider 3-bit numbers and count the number of $1 \rightarrow 0$ flips for each increment. Why does our policy work? As you see from table 2.3, our bank

TABLE 2.3 Bank account

	Bank
000	0
001	1
010	1
011	2
100	1
101	2
110	2
111	3

account has as many tokens as the number of 1 bit. This shows that we have enough tokens in the bank to pay for future $1 \rightarrow 0$ flips.

2.3 Amortized Dictionary

One of the most important structures in computer science is the *dictionary* data structure that supports fast insert and search operations. Here we will discuss a dictionary based on linked lists and sorted arrays. The idea of this data structure is as follows. We will have a linked list of arrays, where array k has size 2^k, and each array has a unique size and is in sorted order. Whether arrays are full or empty is based on the binary representation of the number of items we are storing. For example, with 11 items our dictionary might look like this: $(11 = 1 + 2 + 8)$.

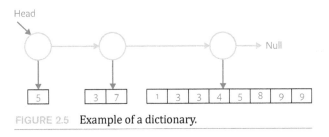

FIGURE 2.5   Example of a dictionary.

In general, we need at most ceiling($\log(n)$) arrays to store n items. How do we insert into this data structure? We create an array of size one and add to the linked list. Since each array must have a different length, insertion requires merging arrays of the same size. As an example, consider inserting 4 into the dictionary in figure 2.5. We get two arrays of size one; thus, we have to merge them. After merging, the dictionary will have two arrays of size two: [4, 5] and [3, 7]. We have to merge them into an array of size four. Figure 2.6 demonstrates the final dictionary.

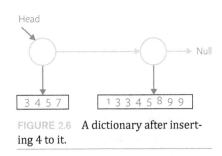

FIGURE 2.6   A dictionary after inserting 4 to it.

In the worst case we may merge all $O(\log(n))$ sorted arrays. Each pair of sorted arrays can be merged in linear time. The total cost model is the following: Creating the initial array of size 1 costs 1, and merging two arrays of size k costs $2k$. The total cost of this insert is $1 + 2 + 4 = 7$. In the general case, the total cost per insert is given by

$$1 + 2 \cdot 2^0 + 2 \cdot 2^1 + 2 \cdot 2^2 + \ldots + 2 \cdot 2^{\log n} = O(n).$$

Therefore, the worst-case runtime complexity of a single insert is $O(n)$. However, on average we do not merge all $O(\log n)$ arrays. What is the amortized cost of insertion? First, we note that each array is a power of 2. Then we observe that adding a new item to the dictionary is equivalent to a bit increment. However, the cost of incrementing is not a constant anymore. Its cost equals to a cost of merging two sorted arrays. It follows that the cost of flipping the k-th bit is 2^k. Consider the least significant bit ($k = 0$). The number of times this bit changes is n, with the cost 2^0. For the next bit ($k = 1$), the cost is 2^1. For the most significant bit ($k = \log(n)$), the cost is $2^{\log n}$. Thus, the total cost of n inserts is given by

$$n \cdot 2^0 + n/2 \cdot 2^1 + n/4 \cdot 2^2 + \ldots + 2 \cdot 2^{\log n} = n + n + \ldots + n = O(n \log n).$$

We have proved that the amortized dictionary data structure has amortized cost $O(\log n)$ per insert.

2.4 Amortized Trees

Recall that a binary search tree is not necessary balanced; therefore, it does not guarantee $O(\log n)$ insertion and searching time that could be in the worst case as bad as $O(n^2)$. There are several ways to make a search tree balanced, though in this section we consider a different approach. Suppose we search a tree multiple times. Don't we want a previous search somehow to affect the next search? Ideally, we want a data structure that adjusts itself to accommodate the observed sequence of operations. The splay tree is a variant of a binary search tree that is designed to do exactly that. The intuition behind splay trees is based on the following observation: If an item was searched once, it is most likely to be searched again. Therefore, the splay tree heuristic is to move a searched item to the root, so that next time the item is searched it would take almost a constant time. Splay trees give up a tree balance in favor of taking advantage of the fact that a large percentage of the searches is caused by only a small subset of data. Splay trees have been introduced by D. Sleater and R. Tarjan in 1985.

The key operation performed on a splay tree is the *splay* operation. `splay(N)` is moving a node N to the root via a sequence of rotations that preserves the binary search tree ordering invariant. Every time a node is accessed in a splay tree, it is moved to the root of the tree. However, splaying is done in a very special way that guarantees $O(\log n)$ amortized bound.

The rotation depends on the positions of the current node N, its parent P, and its grandparent G. There are six types of rotations:

Zig (Zag): A single right (left) rotation. It can only occur when the N node's parent is the root of the tree. This rotation moves the current node N one level up, so N becomes the root.

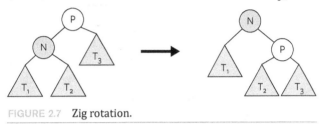

FIGURE 2.7   Zig rotation.

Zig-Zag (Zag-Zig): A double rotation formed by a single Zig (Zag) followed by Zag (Zig) rotation. In the first rotation we rotate N and P. Node N moves a level up and becomes a child of G node. In the next rotation we rotate N and G. Node N moves again a level up, so that nodes G and P become children of N.

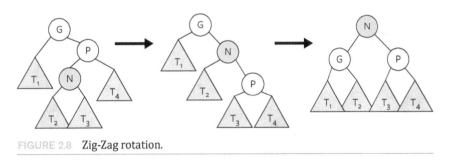

FIGURE 2.8   Zig-Zag rotation.

Zig-Zig (Zag-Zag): A double rotation formed by two single Zig (Zag) rotations in special order. This rotation occurs when N and its parent P are both left (right) children. First, we rotate P and G, and then N and P.

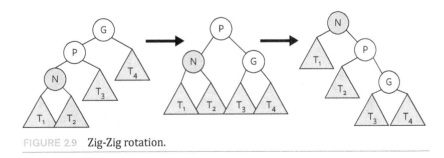

FIGURE 2.9   Zig-Zig rotation.

Note the order of rotations in Zig-Zig rule does matter. Alternatively, we may think to first rotate N and P and then N and G. These rotations will form the tree in figure 2.10. The difference between figures 2.9 and 2.10 might not seem to be that important, but the next example will demonstrate that without this rule, the amortized cost of search is linear.

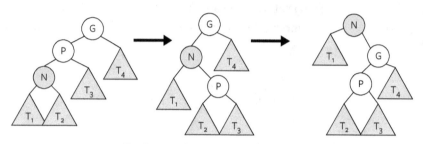

FIGURE 2.10   A wrong Zig-Zig rotation.

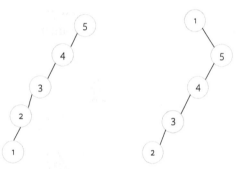

FIGURE 2.11   A linked list of 5 elements.

FIGURE 2.12   The tree after splay(1) has taken place.

Let us consider an example (figure 2.11), where a splay tree is a linked list of ordered elements from 1 to 5. On this tree we will perform the following sequence of operations: splay(1), splay(2), splay(3), splay(4), splay(5). Each splay operation will move a node to the root by performing a chain of single rotations, Zig or Zag. We will use the aggregate method to compute the amortized cost per splay.

To promote node 1 to the root we have performed 4 single Zig rotations.

To promote node 2 to the root we have performed 3 single Zig rotations and one Zag rotation.

To promote node 3 to the root we have performed 2 single Zig rotations and one Zag rotation.

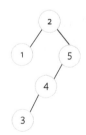

FIGURE 2.13   The tree after splay(1) and splay(2) have taken place.

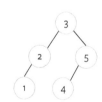

FIGURE 2.14   The tree after splay(1), splay(2), and splay(3) has taken place.

The next two splays, splay(4) and splay(5), will require two and one rotations, respectively. Therefore, the total number of single rotations for 5 splay operations is $4 + 4 + 3 + 2 + 1 = 14$.

Let us generalize this example. Suppose we started with a linked list of ordered items from 1 to n. We run a sequence of n splay operations: splay(1), splay(2), ..., splay(n). Proceeding as in the previous example, the first splay takes $n - 1$ single rotations, splay(2) also takes $n - 1$ single rotations, splay(3) takes $n - 2$ rotations, and so on. The total number of rotations is given by

$$(n-1)+(n-1)+(n-2)+...+2+1=-1+\sum_{k=1}^{n}k=-1+\frac{n(n+1)}{2}=O(n^2).$$

It follows that the amortized cost per splay is $O(n^2)/n = O(n)$. This example demonstrates that in order to achieve $O(\log n)$ amortized bound per splay, we have to have Zig-Zig (Zag-Zag) rotation.

The analysis of running time of splay trees is quite difficult. Any single insert or search might take a linear time in the worst case. But any sequence of m operations on a tree with n nodes takes $O(m \log n)$ time. The proof is far beyond the scope of this book.

REVIEW QUESTIONS

1. What is the definition of the amortized cost using the aggregate method?
2. (**T/F**) Amortized analysis is used to determine the average runtime complexity of an algorithm.
3. (**T/F**) Compared to the worst-case analysis, amortized analysis provides a more accurate upper bound on the performance of an algorithm.
4. (**T/F**) The total amortized cost of a sequence of n operations gives a lower bound on the total actual cost of the sequence.
5. (**T/F**) Amortized constant time for a dynamic array is still guaranteed if we increase the array size by 5%.
6. (**T/F**) If an operation takes $O(1)$ expected time, then it takes $O(1)$ amortized time.
7. Suppose you have a data structure such that a sequence of n operations has an amortized cost of $O(n \log n)$. What could be the highest actual time of a single operation?
8. What is the worst-case runtime complexity of searching in an amortized dictionary?

EXERCISES

1. You have a stack data type, and you need to implement a FIFO queue. The stack has the usual POP and PUSH operations, and the cost of each operation is 1. The FIFO has two operations: ENQUEUE and DEQUEUE. We can implement a FIFO queue using two stacks. What is the amortized cost of ENQUEUE and DEQUEUE operations?

2. We are incrementing a binary counter, where flipping the i-th bit costs $i + 1$. Flipping the lowest-order bit costs $0 + 1 = 1$, the next bit costs $1 + 1 = 2$, the next bit costs $2 + 1 = 3$, and so on. What is the amortized cost per operation for a sequence of n increments, starting from zero?

3. We have argued in the lecture that if the table size is doubled when it's full, then the amortized cost per insert is acceptable. Fred Hacker claims that this consumes too much space. He wants to try to increase the size with every insert by just two over the previous size. What is the amortized cost per insertion in Fred's table?

4. This table supports inserts as well as deletions. The protocol is the following: If an array is full, we double its size on insertion; if an array is 1/4 full, we halve the array size on deletion. Show that the amortized cost of insert and delete is 5.

5. Suppose we perform a sequence of n operations on a data structure in which the i-th operation costs i if i is an exact power of 2 and 1 otherwise. Use aggregate analysis to determine the amortized cost per operation.

6. Suppose we perform a sequence of n operations on a data structure in which the i-th operation costs i if i is an exact power of 4 and 1 otherwise. Use aggregate analysis to determine the amortized cost per operation.

7. A MultiStack data structure has the usual POP and PUSH operations, and the cost of each operation is one unit. Additionally, it has MULTIPOP(k) operation that removes k recently pushed items. If k is bigger than the stack size, it removes all items. We wish to analyze the running time for a sequence of n PUSH, POP, and MULTIPOP operations, starting with an empty stack. What is the worst-case complexity for a sequence of n operations? What is the amortized cost per operation? Use the accounting method.

8. Consider a singly linked list as a dictionary that we always insert at the beginning of the list. Now assume that you may perform any number of insert operations but will only ever perform at most one lookup operation. What is the amortized cost per operation?

Chapter 3

Heaps

H EAPS ARE ONE OF THE MOST important data structures, especially for implementing greedy algorithms using a priority queue. Heaps provide a great option over sorting when input data changes during an algorithm execution. Sorting as we know is a process of arranging elements according to their priorities. However, in many applications we do not need a full sorted order, just the ability to access an element with the highest priority. We start the chapter with classical binary heaps and then extend the definition to amortized heaps that provide the constant amortized cost for insertion and decreaseKey operations. We will be using heaps in a few applications, namely finding the shortest path in graphs, building the minimum spanning tree, and constructing Huffman encoding.

3.1 Binary Heaps

Binary heaps are based on the notion of a complete binary tree. A complete binary tree is a binary tree where each level is completely filled with nodes, except the gap at the bottom level, which is filled from left to right, as illustrated in figure 3.1.

A complete tree with n nodes has a height of floor($\log(n + 1)$). In this example, the tree height is 3. Note that the height of the root is 0.

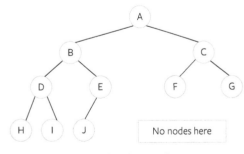

FIGURE 3.1   Example of a complete tree.

A binary heap satisfies one of the following two heap ordering properties:

- *The min-heap property*: The value of each node is greater than or equal to the value of its parent
- *The max-heap property*: The value of each node is less than or equal to the value of its parent

In this course the word *heap* will always refer to a min-heap, unless otherwise noted. Note that a heap may have duplicate elements. To sum up, formally a binary heap can be defined as a collection of items that satisfy the following invariants:

- *Structural property*: States that a heap is a complete tree
- *Ordering property*: The key of the parent node less or equal than the key of children nodes

A heap supports the following operations:

- insert
- deleteMin
- decreaseKey
- build
- meld (merge two heaps)

These operations will be discussed in the subsequent sections. But first, let us discuss heap implementation.

3.1.1 Implementation.

A heap is uniquely represented by storing its data in an array by running a level-order traversal on a tree, with the root at index 1. This allows fast access to each heap element.

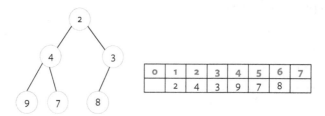

FIGURE 3.2   A heap represented as an array.

Observe that if a node's index is k, its left child is located at $2k$ index, its right child is located at $2k + 1$ index, and its parent is located at $k/2$ index. Array index 0 is left empty to make the indexing work easily.

3.1.2 Insert

The new element is initially appended to the bottom level. If the level is full, we start a new one. In an array-based implementation, we place a new item to the end of the array. This will preserve the structural property.

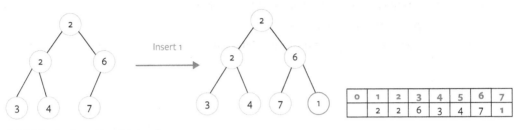

FIGURE 3.3 Inserting 1 into a heap.

At this step, the inserted item may violate the ordering property. We fix this by percolating the item up the tree by swapping positions with the parent, if it's necessary. In figure 3.3 we swap 1 and 6, as shown in figure 3.4.

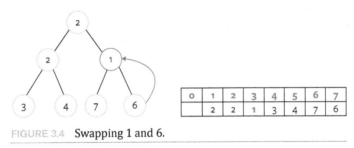

FIGURE 3.4 Swapping 1 and 6.

Again, we observe that new placement of 1 still violates the heap-ordering property. Thus, we swap 1 and 2, as shown in figure 3.5.

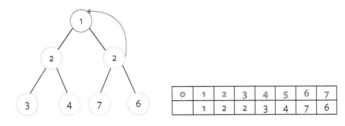

FIGURE 3.5 Swapping 1 and 2.

The worst-case runtime complexity of insertion is $O(\log n)$. This is because a complete tree is a balanced tree and in the worst-case scenario it may require a single swap on each tree level.

3.1.3 DeleteMin

The minimum element can be found at the root of the heap, which is the first element of the array. Clearly, we cannot delete it, since otherwise a tree will be split into two trees. Instead, we move the last element of the heap to the root (this step preserves the structural invariant) and then restore the heap property by percolating it down the tree (this step preserves the ordering invariant). In figure 3.6, we move 8 to the root and then percolate it down by swapping it with the smallest child.

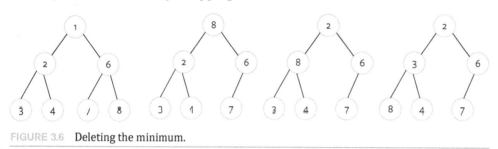

FIGURE 3.6 Deleting the minimum.

This continues until it is less or equal to its children or it reaches the last level. The worst-case runtime complexity of deleteMin is $O(\log n)$, since during percolation it may require a swap on each tree level.

3.1.4 Heapsort

If we run deleteMin n times we will get all heap elements in sorted order. This could be implemented in place by storing the deleted element at the end of the array. Figure 3.7 demonstrates one step of the algorithm; we swap 1 with 8, and then percolate 8 down the tree.

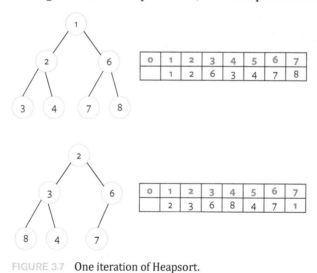

FIGURE 3.7 One iteration of Heapsort.

The worst-case runtime complexity of heapsort is $O(n \log n)$. Heapsort is in place but not stable.

3.1.5   DecreaseKey

In some algorithms we may require changing the key (value) of one of the heap elements. To restore a heap-ordering property, we may need to percolate this item up. The worst-case runtime complexity of decreaseKey is $O(\log n)$. In figure 3.8, we decrease 7 to 2.

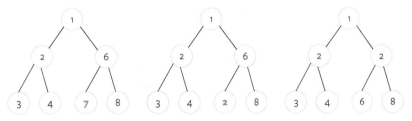

FIGURE 3.8   Demonstration of the decreaseKey operation.

3.1.6   Building a Heap

There are two algorithms to build a heap. The first one is the online algorithm, when the data is not known to us in advance. In this case we build a heap by insertion, starting with an empty array. We will resize the array once it is full. Read about the resizing policy in chapter 2. If we insert n elements, the total cost $T(n)$ is bounded by

$$T(n) = \log 1 + \log 2 + \ldots + \log(n-1) + \log n \leq \log n + \log n + \ldots + \log n + \log n = n \log n.$$

On the other hand (see chapter 1.2 for the proof),

$$T(n) = \log 1 + \log 2 + \ldots + \log(n-1) + \log n = \log(n!) = \Omega(n \log n).$$

Thus, the worst-case runtime complexity of building a heap is $\Theta(n \log n)$.

The second algorithm is offline, when the data is known to us in advance. In this case we can develop a faster algorithm. We will call it "heapify," a process of converting a complete tree into a heap. We begin by placing all the elements into an array in given order. Next, starting at position $n/2$ and working toward position 1, we percolate each element down the tree by swapping it with its smallest child.

Let us consider an example of building a heap on the following set of data: 7, 6, 8, 1, 5, 9, 0, 3, 2, 4. First we place the numbers into a complete tree. This will satisfy the structural invariant.

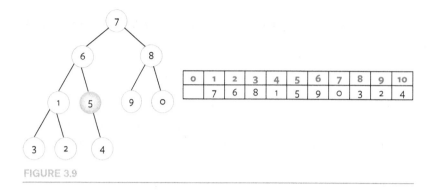

FIGURE 3.9

Then we start at the middle (node 5) and swap it with the child 4. Next, we move to 1. There is nothing to swap for that element, so we move to 8.

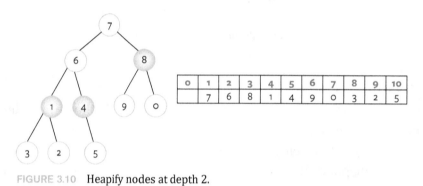

FIGURE 3.10 Heapify nodes at depth 2.

The smallest child of 8 is 0, so we swap 8 with 0 and then move to 6.

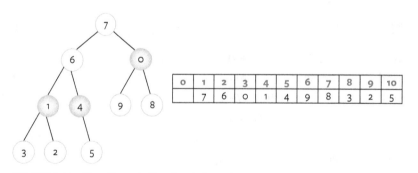

FIGURE 3.11 Heapify node 8 at depth 1.

We swap 6 with the left child 1, and then swap 6 again with the right child 2.

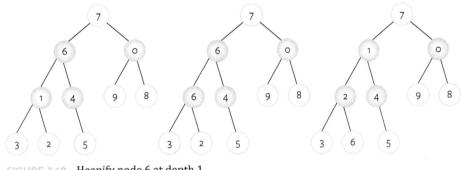

FIGURE 3.12   Heapify node 6 at depth 1.

Finally, we move to the root and swap it with the right child 0. Figure 3.13 shows the final heap.

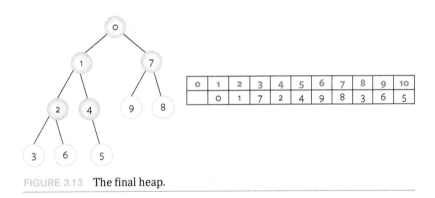

FIGURE 3.13   The final heap.

Now we analyze the worst-case complexity of heapify. During the algorithm execution at most $n/2$, heap elements percolate down the heap. Since the each percolation is $O(\log n)$, the total cost is bounded by $O(n \log n)$. But let us note that not each element was percolated down to a leaf. Thus, we shall derive an asymptotically tight bound. We will count the exact number of swaps (in the worst case) at each level. At the root, we may percolate down h times, where h is the tree height. At a level below, we may have at most $(h - 1)$ swaps. And so on. At the last level, there are zero swaps. Figure 3.14 replicates the number of swaps per level:

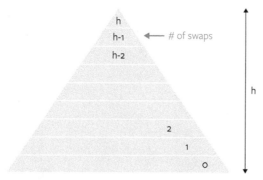

FIGURE 3.14   Demonstrates the number of swaps per level.

Next, we take into account the number of nodes on each level.

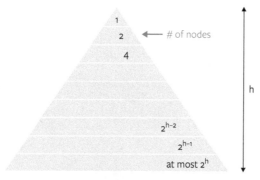

FIGURE 3.15   Demonstrates the number of nodes per level.

We summarize the total number of swaps during the heapification in table 3.1.

TABLE 3.1   The total number of swaps

Height	# of nodes	# of swaps
0	1	h
1	2	$h - 1$
—	—	—
$h - 2$	2^{h-2}	2
$h - 1$	2^{h-1}	1

Finally, we compute the total work by multiplying the number of swaps by the number of nodes on each level. Let $T(n)$ denote the total number of swaps in the worst case. Then, as one can see from table 3.1,

$$T(n) = \sum_{k=1}^{h} k \, 2^{h-k},$$

where $h = \log n$. The finite sum can be further simplified as it follows

$$T(n) = \sum_{k=1}^{h} k \, 2^{h-k} = 2^h \sum_{k=1}^{h} \frac{k}{2^k} \leq 2^h \sum_{k=1}^{\infty} \frac{k}{2^k} = 2^h \, 2 = O(n).$$

This proves that building a heap by running the heapify operation has a linear runtime complexity.

Table 3.2 shows the binary heap operations and their runtime complexities:

TABLE 3.2   Running times for heap operations

Operation	Complexity
findMin	$O(1)$
deleteMin	$O(\log n)$
insert	$O(\log n)$
decreaseKey	$O(\log n)$
buildHeap	$O(n)$

3.2   Binomial Heaps

In the previous section we have proved that insertion takes $O(\log n)$ in the worst case. However, as it easy to see, not all inserts require $\log n$ swaps; some inserts can be performed in constant time. This observation implies that a binary heap *may* exhibit a better amortized complexity of insertion. Let us consider an example of inserting n items in sorted decreasing order from n to 1 into an empty min-heap. We will count the total number of swaps required to insert all the items. For simplicity, we assume that $n = 2^k - 1$. The process of insertion will create a binary heap of height $k - 1 = \log(n + 1) - 1$. Obviously, it takes no swaps to insert the first item n. It takes a single swap to each insert for the next two items $n - 1$ and $n - 2$. Finally, it takes $k - 1$ swaps for each element on the last level (the last level contains 2^{k-1} items.) The total work by inserting n items is given by

$$\sum_{m=0}^{k-1} m 2^m = O(k 2^k) = O(n \log n).$$

Then, the amortized cost per insertion in

$$\frac{O(n\log n)}{n} = O(\log n).$$

This is the same as the worst-case complexity. This simple mathematical computation demonstrates that a binary heap is not suitable for amortized operations and therefore requires creating a different type of heap.

In this section we describe another heap data structure that has a slight improvement in amortized cost over a binary heap. This data structure was introduced by J. Vuillemin in 1978. Each binomial heap is a collection of binomial trees. A binomial tree B_k, of rank k, is defined recursively as follows:

1. B_0 is a single node
2. B_k is formed by joining two B_{k-1} trees

Here are the first four binomial trees:

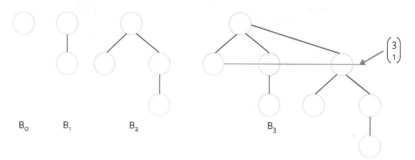

FIGURE 3.16   Example of binomial trees.

The number of nodes on each level l in a binomial tree B_k is defined by binomial coefficients $\binom{k}{l}$, where $0 \le l \le k$. The term *binomial* tree comes exactly from this property. The total number of nodes in B_k is 2^k, as it follows from

$$\binom{k}{0} + \binom{k}{1} + \dots + \binom{k}{k-1} + \binom{k}{k} = 2^k$$

Another interesting property of a binomial tree B_k is that when we remove the root, the tree will break into k binomial trees $B_0, B_1, \dots, B_{k-1}$. Figure 3.17 shows a new way of looking at B_3.

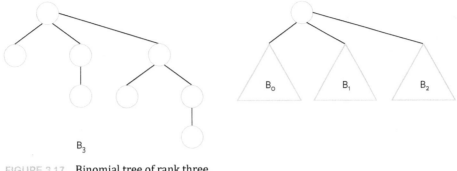

B_3

FIGURE 3.17   Binomial tree of rank three.

A binomial heap is a collection (a linked list) of at most *ceiling*(log n) binomial trees in increasing order of size, where each tree has a heap ordering property. In a binomial heap there is at most one binomial tree of any given rank. In order to have constant time access to the top element, we store the pointer to the smallest root. Figure 3.18 demonstrates a binomial heap of 13 elements:

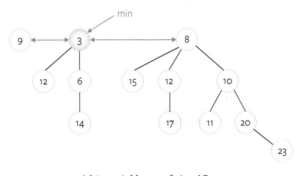

FIGURE 3.18   A binomial heap of size 13.

Observe that the number of elements that can be stored in a heap relates to its binary expansion. To store 13 elements, we need B_0, B_2, and B_3 binomial trees. This is due to the binary expansion $13_{10} = 1101_2$. If we need to store 25 items, the heap will be a collection of B_0, B_3, and B_4, since $25_{10} = 11001_2$. Thus, a binomial heap with n nodes has number of binomial trees equal to the number of 1's bits in binary representation of n. Having this in mind, we always will assume in the worst-case analysis that there are $O(\log n)$ binomial trees in a binomial heap with n nodes.

3.2.1   Merging

Binomial heaps allow faster merging, compared to binary heaps. Note, binary heaps are complete binary trees, and two complete binary trees cannot easily be linked to one another. Consider the merging of two binomial heaps on the following example in figure 3.19.

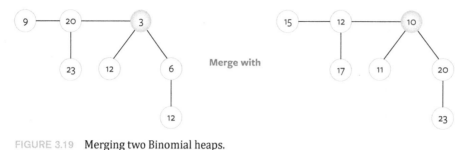

FIGURE 3.19   Merging two Binomial heaps.

First we merge two heaps as we merge two linked lists; it takes $O(1)$ time. We get the heap shown in figure 3.20.

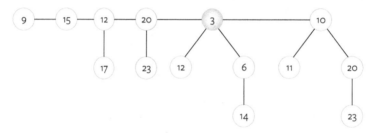

FIGURE 3.20   The result of joining two top linked lists.

This heap is not a binomial heap yet, since it has trees of the same ranks. Thus, we need to combine binomial trees of the same rank. This can be done by making the smaller root the child of the larger root. It also takes $O(1)$ time; however, it may require to merge $O(\log n)$ trees in total. Thus, the worst-case runtime complexity of merging is $O(\log n)$. In our example we need to merge two trees of rank 0 and two trees of rank 1.

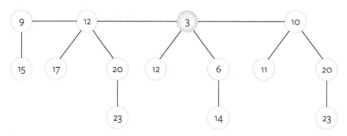

FIGURE 3.21   The result of merging two trees of rank 0 and two trees of rank 1.

Finally, we combine two trees of rank 2 to get the heap in figure 3.22.

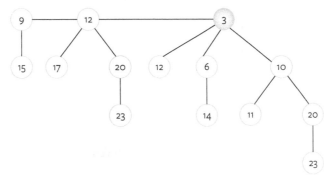

FIGURE 3.22   The result of merging two trees of rank 2.

In conclusion, we note that merging two binomial heaps is related to a binary addition. In the previous example, the result of merging B0B1B2 with B0B1B2 is a binomial heap B1B2B3. This can be viewed as a binary addition

$$
\begin{array}{r}
111 \\
111 \\
\hline
1110
\end{array}
$$

where the result of addition 1110 translates into a heap $B_1B_2B_3$.

3.2.2.   DeleteMin

The algorithm is as follows:

1. Find the binomial tree that contains the minimum element
2. Delete the root and move all subtrees to the top list
3. Merge the binomial trees of the same rank
4. Set a pointer to the new minimum

Note that deleting the root of B_k results in $B_0, B_1, ..., B_{k-1}$ binomial trees. It follows that the worst-case complexity of deleteMin is $O(\log n)$, which is the same as merging two heaps. Let us execute the algorithm on the heap in figure 3.23.

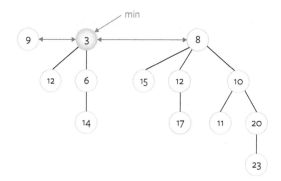

FIGURE 3.23　We will perform deleteMin on this heap.

After deleting the minimum, the heap transforms into what is shown in figure 3.24.

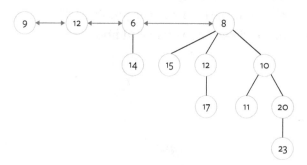

FIGURE 3.24　The result of deleting the min.

Next, we merge two B_0 and B_1 to get what is shown in figure 3.25.

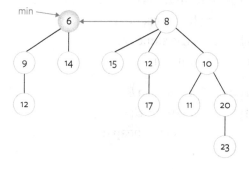

FIGURE 3.25　The result of merging B_0 and B_1.

3.2.3. Insert

Essentially, insertion is merging two heaps, one of size 1 and the other of size n. Therefore, it takes $O(\log n)$ in the worst case. This case occurs only if a binomial heap of size $n = 2^m - 1$ contains binomial trees of all orders, namely the following trees B_0, B_1, ..., B_{m-1}. Then, inserting a new item into this heap will case m binomial trees to merge. The first merge will result in a heap $B_1 B_1 B_2 ... B_{m-1}$. The second merge will result in a heap $B_2 B_2 B_3 ... B_{m-1}$. And so on. After $m = O(\log n)$ merges we will get a binomial heap that contains only a single binomial tree of order m.

It should be clear that not each insertion requires merging all $O(\log n)$ binomial trees. Let us compute amortized cost per insertion. We will use the accounting method. We show that assigning two tokens to a single insert is sufficient. Here is our assignment: One token is paid for creating a single binomial tree, and the other is for future tree merging. In this model each binomial tree in a heap has a single token associated with it. When we merge two trees of the same rank, we use one token to pay for merging and keep the second token for the next merger (if it will ever be required.) It follows that single insertion into a binomial heap has a constant amortized cost.

Here is another way to prove that amortized cost of a single insert is constant. Recall that a binomial heap of size n is associated with a binary expansion of n. When we insert a new item into it, we merge two heaps, one of size 1 and the other of size n. This is equivalent to a binary addition, namely incrementing a binary representation of n. We have proved in chapter 2 that amortized cost of binary increment is $O(2)$.

3.2.4 Building a Binomial Heap

We have studied two algorithms of building a binary heap. One is an offline algorithm (building by insertion) with the runtime complexity $O(n \log n)$; the other is an online algorithm (building by heapifying) with the runtime complexity $O(n)$. The cost of building a binomial heap of n elements by insertion is $O(n)$, even if the data is not known to us in advance.

Finally, we summarize runtime complexities of binary and binomial heaps in table 3.3 (here, "ac" stands for amortized cost).

TABLE 3.3 Running times for heap operations

	Binary	Binomial
findMin	$O(1)$	$O(1)$
deleteMin	$O(\log n)$	$O(\log n)$
insert	$O(\log n)$	$O(1)$ (ac)
decreaseKey	$O(\log n)$	$O(\log n)$
merge	$O(n)$	$O(\log n)$

3.3   Fibonacci Heaps

The Fibonacci heap data structure was invented by Fredman and Tarjan in 1987. The general idea is to have a more relaxed structure (compared to binomial heaps) that will improve `decreaseKey` complexity to constant amortized time. The trees in a Fibonacci heap are not constrained to be binomial trees. Figure 3.26 shows an example of a Fibonacci heap.

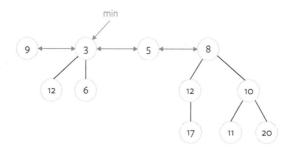

FIGURE 3.26   Example of a Fibonacci heap.

The high-level idea of a `decreaseKey` algorithm is to take the node you want to decrease, change its value, disconnect it and its entire subtree from where it is, and attach it to the tree root list. This is clearly $O(1)$ time. This attractive feature of Fibonacci heaps allows a performance improvement to many algorithms, in particular, the Dijkstra's shortest path algorithm, (see chapter 4.5.1.2) bringing its runtime complexity to $O(E + V \log V)$.

Let us discuss a bird's-eye view of a `decreaseKey` algorithm on the heap in figure 3.27.

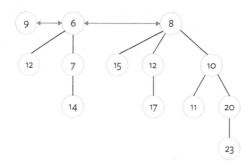

FIGURE 3.27   We will perform decreasekey on this heap.

Suppose we want to change 7 to 5. Since 7 is not the root of the tree, its value decrementing will break the heap order. We cut the tree rooted at 7 from its parent and move it to the top level.

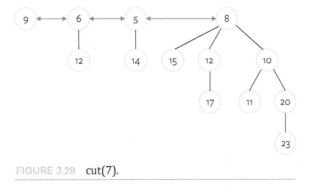

FIGURE 3.28 cut(7).

Running a new function cut(7), we are guaranteed that changing its value to 5 surely does not break the heap order. However, we may end up with having extremely sparse trees of high ranks as well as with several trees of the same rank. In order to avoid this problem, we limit the number of cuts among the children of any vertex to two. This is done by implementing another function marked(v) that keeps a track of cuts of all children of v. Clearly, after a call to decreaseKey we won't have a binomial tree anymore. We will fix the binomial heap property when deleteMin is called. However, the problem is that we can no longer prove the bound on the time of deleteMin. That heap may contain more than $O(\log n)$ binomial trees, and some of them are not necessary binomial. There's a clever way to fix this by implementing "cascading cuts." The algorithm was designed by M. Fredman and R. Tarjan. The algorithm is beyond the scope of this course.

Fibonacci heaps have another advantage: The worst-case time complexity of the insert is $O(1)$. How do we insert into the Fibonacci heap? Just add a single node to the top level. Do not merge binomial trees! We may have several trees of the same rank. We will fix the heap when deleteMin is called. Clearly, lazy insertion runs in $O(1)$ time in the worst case.

In table 3.4, we summarize runtime complexities of different heaps. There "ac" stands for the amortized time complexity.

TABLE 3.4 Running times for heap operations

	Binary	Binomial	Fibonacci
findMin	$\Theta(1)$	$\Theta(1)$	$\Theta(1)$
deleteMin	$\Theta(\log n)$	$\Theta(\log n)$	$O(\log n)$ (ac)
insert	$\Theta(\log n)$	$\Theta(1)$ (ac)	$\Theta(1)$
decreaseKey	$\Theta(\log n)$	$\Theta(\log n)$	$\Theta(1)$ (ac)
merge	$\Theta(n)$	$\Theta(\log n)$	$\Theta(1)$ (ac)

1. What is the worst-case runtime complexity of finding the smallest item in a binary min-heap?

2. What is the worst-case runtime complexity of finding the largest item in a binary min-heap?

3. How many binomial trees does a binomial heap with 31 elements contain?

4. How many binomial trees are in a binomial heap of size n?

5. What is the worst-case runtime complexity of inserting into a binomial heap?

6. What is the worst-case runtime complexity of searching in a binomial heap?

7. What is the amortized cost of inserting into a binomial heap?

8. What is the worst-case runtime complexity of deleteMin() from a binomial heap?

9. (**T/F**) The following array is a max heap: [10, 3, 5, 1, 4, 2].

10. (**T/F**) In a binary max-heap with n elements, the worst-case runtime complexity of finding the second largest element is $O(1)$.

11. (**T/F**) If item A is an ancestor of item B in a heap then it must be the case that the insert(A) operation occurred before insert(B).

12. (**T/F**) Using a binary heap we can sort any array of size n in $O(n)$ time.

13. (**T/F**) In a binomial min-heap with n elements, the worst-case runtime complexity of finding the smallest element is $O(1)$.

14. (**T/F**) In a binomial min-heap with n elements, the worst-case runtime complexity of finding the second smallest element is $O(1)$.

15. (**T/F**) By using a binomial heap we can sort data of size n in $O(n)$ time.

16. (**T/F**) Given a Fibonacci heap of size n, the maximum number of trees is that heap is n.

EXERCISES

1. Given a sequence of numbers, 3, 5, 2, 8, 1, 5, 2,

   a. draw a binary min-heap (in an array form) by inserting these numbers, reading them from left to right; and

   b. show an array that would be the result after the call to deleteMin() on this heap.

2. Devise an algorithm of merging two binary heaps. What is its runtime complexity?

3. Suppose you have two binary min-heaps, A and B, with a total of n elements between them. You want to discover if A and B have a key in common. Devise an algorithm to this problem that takes $O(n \log n)$ time.

4. The values 1, 2, 3, ..., 63 are all inserted (in any order) into an initially empty min-heap. What is the smallest number that could be a leaf node?

5. Prove that it is impossible construct a min-heap (not necessarily binary) in a comparison-based model with *both* the following properties:

   a. deleteMin() runs in $O(1)$

   b. buildHeap() runs in $O(n)$, where n is the input size

6. Given an unsorted array of size n, devise a heap-based algorithm that finds the k-th largest element in the array. What is its runtime complexity?

7. Recall that two sorted arrays of size n can be merged into a single sorted list in linear time $O(n)$. Suppose there are $k > 2$ sorted arrays, each of size n. Devise a heap-based algorithm that merges k arrays and requires at most $O(k)$ extra space.

8. Given a stream of data (its size is unknown in advance), devise a heap-based algorithm that finds the k-th largest element in the array. Your algorithm must take at most $O(k)$ extra space. What is its runtime complexity?

9. Given a stream of data (its size is unknown in advance), devise a heap-based algorithm that finds the median of elements read so far. What is its runtime complexity?

10. Given a sequence of numbers, 3, 5, 2, 8, 1, 5, 2, 7,

    a. draw a binomial heap by inserting these numbers, reading them from left to right; and

    b. show a heap that would be the result after the call to deleteMin() on this heap.

11. Discuss the relationship between inserting into a binomial heap and binary increment.

12. Discuss the relationship between merging two binomial heaps and adding two binary numbers.

13. Discuss the relationship between inserting into a binomial heap and a Fibonacci heap.

14. Devise an algorithm of deleting any item from a binomial heap. What is its runtime complexity?

15. Devise an algorithm to find all nodes less than some given value X in a binomial heap. Analyze its complexity.

Chapter 4

Greedy Algorithms

G REEDY ALGORITHMS DO NOT HAVE A formal definition, but all of them possess the following characteristics:

- They make a sequence of choices.

- Each choice is the best available at each step.

- Earlier decisions made during execution are never undone.

- They do not always yield the optimal solution.

Greedy algorithms have several advantages over other algorithmic approaches. The first one is simplicity: Greedy algorithms are often easier to describe and implement. The second is efficiency: The greedy approach can often produce more efficient solutions. At the same time, they have a drawback: Showing that a greedy algorithm is correct often requires a non-trivial proof.

How can we tell if a greedy approach will solve a particular problem? There is no guarantee that such a greedy algorithm exists; however, a problem to be solved must obey the following two common properties:

- Optimal substructure

- Greedy-choice property

An optimal substructure means that an optimal solution to the original problem contains optimal solutions to all of its subproblems. The proof of optimal substructure correctness is usually by induction.

A greedy-choice property means that a globally optimal solution is obtained by making a locally optimal (greedy) choice. This choice is made to solve each subproblem and may depend on choices that have been made to date, but it cannot depend on any future choices. The proof that a greedy choice for each subproblem yields a globally optimal solution is usually by contradiction.

Where does greedy approach efficiency come from? A greedy algorithm can be described as a multistage decision-making process, and therefore we can construct a tree to enumerate all possible decisions. During the algorithm execution, we don't consider all available choices at any given node, but use a greedy heuristic to pick just one, the highest-ranking child. In this model a greedy technique can be viewed as finding a set of paths from the root to a leaf node. Consider a board game where two players alternately take turns. We can use a tree to represent all possible moves until the game ends. Each node corresponds to a position, and each edge corresponds to a move. In this game tree, the number of nodes on each level is exponential in the tree height. Thus, a brute-force algorithm will have an exponential (in height) runtime complexity. A greedy algorithm, in contrast, will make only a single greedy choice at each tree level; therefore, its runtime complexity will be proportional to the height.

We conclude the introduction with a few remarks on implementation. In order to make greedy choices efficiently we have to use a certain data structure. One simple choice is an unsorted array. The better choice may be a priority queue that allows accessing the highest-ranking choice in constant time. Alternatively, we could use a sorted array, though this may be more expensive compared to heaps. In this chapter we start with two greedy algorithms when a single sorting is sufficient and then proceed to other algorithms when use of a priority queue is advantageous.

4.1   The Money Changing Problem

In this problem we are to compute the minimum number of coins needed to make change for a given amount m and given set of n denominations. Assume that we have an unlimited supply of coins. As an example, let us use US currency (pennies, nickels, dimes, and quarters) and the amount to change is $m = \$0.40$. There are several ways to make change, $0.40 = 4*0.10$ (four dimes) or $0.40 = 2*0.10 + 4*0.05$ (two dimes and four nickels). But intuitively we can get the smaller number of coins if we start with the largest coin first: $0.40 = 0.25 + 0.10 + 0.05$. This suggests the following greedy algorithm to make change: Start with the largest coin and use it as many times as possible; then use the second largest coin, and so on. Will we get the least number of coins? We prove the algorithm correct by contradiction. If we do not choose the largest coin, is there a better solution? Assume that our algorithm does not take the largest coin (which is a quarter). Then we will need a combination of smaller coins (pennies, nickels, or dimes) to add up to a quarter. It follows that the change won't be optimal, since we will end up with more coins.

Let us discuss the algorithm efficiency. We will visualize the algorithm as a quad choice tree (figure 4.1) where each vertex contains an amount to change and each edge is a denomination.

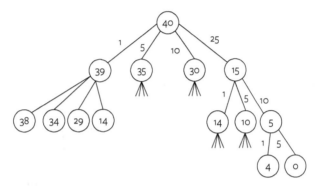

FIGURE 4.1   A choice tree to change 40 cents using pennies, nickels, dimes and quarters.

In the brute-force approach we will have to try each available denomination. This will lead to $O(4^h)$ runtime, where h is the tree height. In the greedy approach, we will always choose the largest available denomination. It is clear that the complexity of this approach is $O(h)$. Here we assumed that we could get the largest coin in the constant time. We can always do that by sorting all denominations in descending order and then traversing them in that order. In a general case, when we have a set of n denominations, this choice tree has height $h = n$; therefore, the algorithm runtime complexity is $O(n)$.

Lastly, we demonstrate on the example that a greedy choice does not necessarily yield the optimal solution. Let us imagine a different denomination system, where in addition to pennies, nickels, dimes, and quarters we have a 20-cent coin. Then, running a greedy approach, we still get three coins: $0.40 = 0.25 + 0.10 + 0.05$. However, the optimal solution contains only two coins: $0.40 = 2*0.20$. This example emphasizes an importance of proving the algorithm correctness.

4.2   Scheduling Problem

There is a set of n requests. Each i-th request has a starting time $s(i)$ and finish time $f(i)$. Assume that all requests are equally important and $s(i) \leq f(i)$. Our goal is to develop a greedy algorithm that finds the largest compatible (non-overlapping) subset of requests. This problem is interesting because among many available greedy strategies it is not obvious which one to choose. One approach is to sort requests with respect to $s(i)$ in ascending order. This one is not going to work (see figure 4.2). In that example the solution will consist of one request.

FIGURE 4.2   The earliest starting time strategy.

You may think that starting with the shortest $f(i) - s(i)$ request first will be the right strategy. See figure 4.3 for the counterexample.

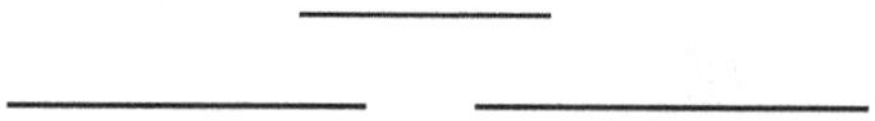

FIGURE 4.3   The shortest interval strategy.

Another possible strategy is to take into consideration the number of overlapping intervals. In this strategy we start with an interval that has the smallest number of overlaps with other intervals. See figure 4.4 for the counterexample. It demonstrates that using this strategy we get only three intervals; however, the optimal solution has four intervals.

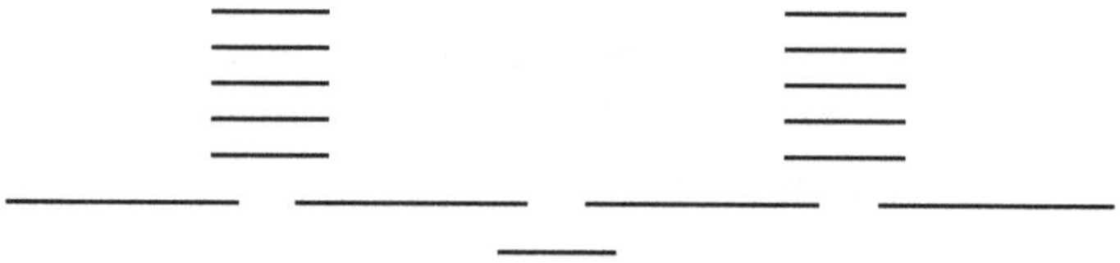

FIGURE 4.4   The smallest number of overlaps.

Finally, we consider a strategy of taking intervals with respect to finish time $f(i)$, a request with the earliest finish time first. In this approach we sort requests with respect to $f(i)$ in ascending order. Pick a request that has the earliest finish time. Delete all requests that overlap with it. Repeat.

The running time is $O(n \log n)$ for sorting plus $O(n)$ for the greedy collection of activities. Does it always find an optimum?

4.2.1   Proof of Optimality

We assume that all intervals are sorted with respect to the finish time. Let $\{i_1, i_2, ..., i_k\}$ be a subset of intervals chosen by our greedy algorithm and $\{j_1, j_2, ..., j_m\}$ be the optimal subset of intervals. We will prove by induction that $f(i_r) \leq f(j_r)$ for $\forall r \leq k$ and thus our solution cannot be worse than the optimal one.

> *Base case*: $r = 1$. This is true, $f(i_1) = f(j_1)$, because we start with the earliest finish time.

> *Inductive hypothesis*: Let us assume $f(i_{r-1}) \leq f(j_{r-1})$.

> *Inductive step*: We need to prove $f(i_r) \leq f(j_r)$.

Note, $f(j_{r-1}) \leq s(j_r)$ since in the solution the intervals cannot overlap. Thus, using the inductive hypothesis $f(i_{r-1}) \leq f(j_{r-1}) \leq s(j_r)$ we arrive at $f(i_{r-1}) \leq s(j_r)$. Since j_r is in the optimal set, then in the next step our greedy algorithm must pick the j_r interval. It follows $f(i_r) \leq f(j_r)$.

Next, we need to prove that our solution $\{i_1, i_2, ..., i_k\}$ has the same size as the optimal solution (i.e., $k = m$). We prove this by contradiction. Let us assume that $k < m$. There must be a request j_{k+1} such that $f(j_k) \leq s(j_{k+1})$ and $f(i_k) \leq f(j_k)$. Combining these two inequalities, it follows $f(i_k) \leq s(j_{k+1})$. This means that a request j_{k+1} does not overlap with any $i_1, i_2, ..., i_k$ requests. So, our greedy algorithm would not stop at i_k and choose j_{k+1} as the next request. Contradiction, the size of our solution, is bigger than k.

4.3 Huffman Code

In 1948 Claude Shannon established that there is a fundamental limit to lossless data compression. This limit is called the entropy rate H. Entropy is a measure of the amount of information contained in the source. It is possible to compress the source, in a lossless manner, with a compression rate close to the entropy H. But it is mathematically impossible to do better than H. The ASCII table is the simplest example of data compression. In that model we assign a fixed number of bits (called a codeword) to a character, namely 8 bits. But we can achieve a better compression ratio if we assign a variable number of bits to each character. It is known (statistically) that the character "e" is much more likely to appear than "u." In this model each letter in the alphabet of size n has a certain probability p_k. We can define probabilities by counting frequencies of each character in the input text. The entropy H is given by

$$H = \sum_{k=0}^{n} p_k \log \frac{1}{p_k}.$$

H is the lower bound on the average number of bits to code a character. Using standard distribution of characters in the English language, we get $H = 4.07$ bits/char.

This theoretical result cannot be directly used to compute the number of bits per letter in a data compression algorithm, due to the fact that the log values are not integers. If we round up the logs, then the solution is not an optimal. In 1952 David Huffman developed a greedy algorithm to assign a prefix-free codeword to each character in the text according to their frequencies. A prefix-free code is one where no codeword is a prefix of another codeword.

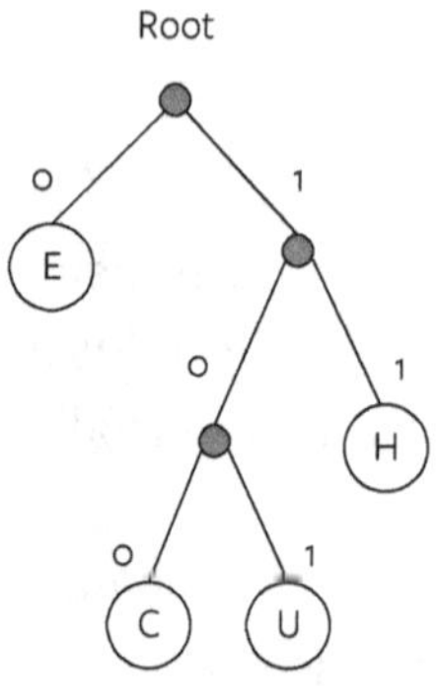

We will be using a full tree to map each character to a binary string. A codeword is a path from the root to the character. In figure 4.5 a codeword for C is 100 and a codeword for H is 11. Using a prefix-free code it is easy to encode and decode data. To encode, we need only to concatenate the codewords for each character. To decode, we scan the text from left to right, and as soon as we recognize a codeword, we print the corresponding character.

FIGURE 4.5 Prefix-free binary codes.

In general, we want to minimize the overall length of encoding, namely

$$\text{cost}(T) = \min \sum_{k=1}^{n} f(x_k)\, d(x_k)$$

where $f(x)$ is a frequency of x_k character and $d(x_k)$ is a depth of x_k in the tree T. This suggests a greedy approach to constructing a tree. We need to put characters with the lowest frequencies to the bottom of a tree. This will guarantee longer binary strings assigned to them. Characters with the high frequency should be at the top, so they will have shorter codewords. Such a tree is called a Huffman tree.

4.3.1 Example: Building a Huffman tree

Let us draw a Huffman tree for the following table of frequencies:

TABLE 4.1 A table of frequencies

char	A	M	L	E	K	B	U	X
freq	34	21	14	13	11	9	8	7

Initially, there are only single-node trees: one for each character.

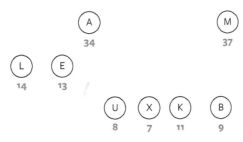

FIGURE 4.6   Single-node trees.

Next we select two characters of the smallest frequencies (they are U and X) and form a new parent node with the frequency $8 + 7 = 15$, and connect it to U and X.

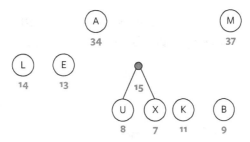

FIGURE 4.7   A parent node for U and X.

Once two nodes in a tree are connected, they are removed from consideration. However, their parent node is still in the game. Again, select two characters of the smallest frequencies (they are K and B), form a new node with the frequency $11 + 9 = 20$, and connect it to K and B. In the next step we connect L and E. The result is depicted in figure 4.8.

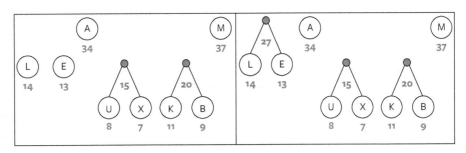

FIGURE 4.8   The result of joining K and B, and then L and E.

Next, we join 15 and 20, and then 27 and 34 as in figure 4.9

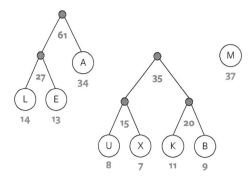

FIGURE 4.9 The result of joining 15 and 20, and then 27 and 34.

Continue connecting nodes until there is only one tree left. That tree is the optimal Huffman coding tree. Lastly, we assign 0's and 1's to the edges.

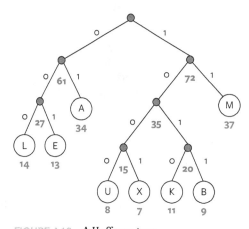

FIGURE 4.10 A Huffman tree.

Table 4.2 is a table of codewords.

TABLE 4.2 A table of codewords

char	A	M	L	E	K	B	U	X
freq	34	37	14	13	11	9	8	7
codeword	01	11	000	001	1010	1011	1000	1001

To get the total number of bits needed to compress a text (given the frequency table) we multiply the frequency of each character by the codeword length in bits:

$$34*2 + 37*2 + 14*3 + 13*3 + 11*4 + 9*4 + 8*4 + 7*4 = 363.$$

4.3.2 Proof of Optimality

We will prove it by induction on the number of characters.

Base case: Two characters. The tree is unique; therefore, it is optimal.

Inductive hypothesis: Assume that a Huffman tree of any $n - 1$ characters is optimal.

Inductive step: We need to prove that a Huffman tree of n characters is optimal.

Given a set A of n characters x_k with some frequencies $f(x_k)$, where $k = 1, 2, ..., n$. If we run our greedy algorithm we will get a tree T over A. We will prove that T is optimal in a sense that T minimizes the overall length of encoding:

$$\text{cost}(T) = \min \sum_{k=1}^{n} f(x_k) \, d(x_k),$$

where $d(x_k)$ is a depth of x_k in the tree T. Note that $d(x_k)$ is the number of bits of a codeword associated with x_k. Let us run a single step of our greedy algorithm. We choose two characters, x_1 and x_2, with the lowest frequencies $f(x_1)$ and $f(x_2)$. Then we join them to create a parent node x^* with a frequency $f(x^*) = f(x_1) + f(x_2)$. After this step the number of characters in consideration is decreased by one (we removed x_1 and x_2 and added x^*). Let us call this new set of characters by A^*:

$$A^* = A \setminus \{x_1, x_2\} \cap \{x^*\}.$$

Since the size of A^* is $n - 1$, we can apply the inductive hypothesis and thus build an optimal Huffman tree over A^*. We will call this tree by T^*. Let us summarize the construction; we say that we build a tree T (over a set A) by running one step of the algorithm for two characters and then using inductive hypothesis for the rest of characters (a set A^*):

$$\text{cost}(T) = \sum_{k=1}^{n} f(x_k) \, d(x_k) = f(x_1) \, d(x_1) + f(x_2) \, d(x_2) + \sum_{k=1}^{n-2} f(x_k) d(x_k).$$

By construction the cost of T is related to cost (T^*) as follows:

$$\text{cost}(T) = f(x_1) d(x_1) + f(x_2) d(x_2) + \text{cost}(T^*) - f(x^*) d(x^*).$$

Here, $f(x^*) = f(x_1) + f(x_2)$ and $d(x_1) = d(x^*) + 1$ and $d(x_2) = d(x^*) + 1$. Substituting these into the previous equation, we get

$$\text{cost}(T) = \text{cost}(T^*) + f(x_1) + f(x_2).$$

It is important to observe that we can reverse the construction process; namely, we can get an optimal tree T^* from T by removing two lowest frequency nodes if they are siblings (see 4.3.2.1 for details)

$$\text{cost}(T^*) = \text{cost}(T) - f(x_1) - f(x_2)$$

Having this in mind we proceed to the next step. We prove optimality of T by contradiction. Assume that there is another tree T_1 over the set A such that cost $(T_1) < $ cost (T). For the tree T_1 we can perform the same reverse process to get another optimal tree T_1^*:

$$\text{cost}(T_1^*) = \text{cost}(T_1) - f(x_1) - f(x_2).$$

Since $\text{cost}(T_1) < \text{cost}(T)$, it follows $\text{cost}(T_1^*) < \text{cost}(T^*)$. But tree T^* is optimal. This is a contradiction. ■

4.3.2.1   What if x_1 and x_2 are not siblings?

Lemma. *Let x and y be two characters such that $f(x)$ and $f(y)$ are minimal. Then there is an optimal prefix code such that x and y are siblings.*

Proof. Let T be the optimal tree and z be a sibling of x such that $d(x) = d(z) \geq d(y)$. Consider two cases.

Case 1. $d(z) = d(y)$. If z and y are at the same depth, we can swap them. The cost of the optimal tree T won't change.

Case 2. $d(z) > d(y)$. Since z is located deeper within the tree T than $f(z) < f(y)$, we swap z and y and call that tree T_1 as in figure 4.11.

Next, we compute the cost of T:

$$\text{cost}(T) = f(x)d(x) + f(z)d(x) + f(y)d(y) + \ldots$$

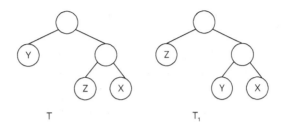

FIGURE 4.11   The result of swapping z and y.

the cost of T_1

$$cost(T_1) = f(x)d(x) + f(z)d(y) + f(y)d(x) + \ldots$$

and subtract them to get

$$cost(T_1) - cost(T) = f(z)d(y) + f(y)d(x) - f(z)d(x) - f(y)d(y) = (f(y) - f(z))(d(x) - d(y)).$$

Since $f(y) > f(z)$ and $d(x) < d(y)$, it follows that $cost(T_1) \leq cost(T)$. But T is the optimal tree, thus $cost(T_1) = cost(T)$. ∎

4.3.3   Runtime Complexity of Building a Huffman Tree

Let us assume that a frequency table is given to us and its size is n. The algorithm works by repeatedly connecting a pair of nodes that have the smallest frequencies. The frequency of the new node is the sum of the frequencies of the connected nodes. We keep (node, frequency) in a min-heap. In each step of the algorithm we extract two nodes with the smallest frequencies, create a new parent node, and insert it back into the heap. The whole process takes $O(n \log n)$ time. Observe that using an unsorted array instead of a min-heap is less efficient. It will take us $O(n)$ to find the minimum, and $O(1)$ to insert a parent node. This will lead to $O(n^2)$ time.

4.3.4   Storing a Huffman Tree

In this section we will discuss decompression—a process of translating the stream of prefix codes back to the characters. It should be clear that in order to decompress an encoded text file we have to have the same Huffman tree that was used to compress. Therefore, every compressed file must have the whole Huffman tree stored in a binary form. How to store a full tree? It turns out a single bit per node is sufficient: a 0 bit for an internal node and a 1 bit for a leaf. We output nodes in preorder traversal. Once we hit a

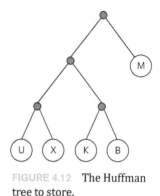

FIGURE 4.12   The Huffman tree to store.

leaf we output a binary ASCII code (8 bits) for that character. As an example, consider the tree in figure 4.12

The preorder traversal yields 0001U1X01K1B1M. The ASCII code for character U is 85, or 01010101 in binary. Taking into account ASCII codes for all other characters, we get the following encoded string: 0001010101011010110000101001 01110100001010100110. The total number of bits required to store that tree is $9 + 8*5 = 49$.

The Huffman tree is always stored at the beginning of a compressed file and is called a header. During decompression we read the header and restore the Huffman tree recursively from a preorder traversal.

4.4   Minimum Spanning Trees

Given a weighted undirected connected graph $G = (V, E)$, a spanning tree is a subgraph of G that contains all vertices and it is a tree. The cost of a tree is the sum of the weights of its edges. A minimum spanning tree (MST) is a spanning tree with the minimum cost.

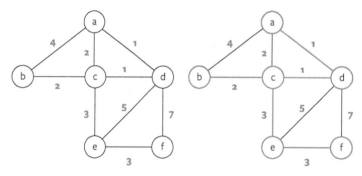

FIGURE 4.13   A graph on the left and its MST (blue edges) on the right.

It is important to note that a given graph is undirected! For directed graphs an MST problem is defined in a different way and called an arborescence problem (which we won't cover in this text). Before we proceed with a greedy approach, let us discuss a brute-force approach. We find all spanning trees (using a BFS, for example) for a given graph and then choose the one with the smallest cost. It turns out this approach is quite expensive. In 1889 Arthur Cayley proved that the number of spanning trees in K_n (a complete graph on n vertices) is n^{n-2}, and thus the brute force approach has an exponential runtime complexity. We will omit a proof of Cayley's theorem, and instead we will discuss a polynomial time algorithm for finding an MST.

4.4.1   Prim's Algorithm (1957)

For any weighted undirected graph $G = (V, E)$, the algorithm builds a minimum spanning tree T one vertex at a time. Here are the algorithm steps:

1. Start with an arbitrary vertex and add it to an empty tree T. This vertex will be the root of T.
2. Expand T by adding a vertex from $V\backslash T$, having the minimum weight edge and having exactly one end point in T.
3. Update distances from all vertices in T to adjacent vertices in $V\backslash T$.
4. Continue to grow the tree until T gets all vertices, $T = V$.

Step 2 is a greedy choice: Among all adjacent vertices to T we pick the one that has the minimum weight edge. Step 3 is the most important step; we update only the shorter edges from T to $V\backslash T$. The greedy choice implies that we have to use an intermediate data structure, which will allow us to find a vertex with the shortest edge in the most efficient way. This suggests use of a priority queue.

4.4.1.1   Example: Building an MST

Let us run Prim's algorithm on the graph in figure 4.14. We will keep a binary min-heap H as an intermediate data structure. Every element of the min-heap contains a vertex number and a key value of the vertex, which is an edge weight from a tree T to the vertex. In min-heap H we won't show vertices that are not yet connected to T by an edge.

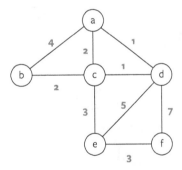

FIGURE 4.14   The Prim algorithm illustrated on this graph.

We start at vertex a, so $T = \{a\}$. Edge weights from T to all adjacent vertices are $|ab| = 4$, $|ac| = 2$, $|ad| = 1$. We update adjacent vertices in a heap $H = \{d_1, c_2, b_4\}$, where the subscripts denote an edge weight from a tree T to the vertex. We pick the shortest vertex in H (which is d) and add it to T. After this step $T = \{a, d\}$ and $H = \{c_2, b_4\}$.

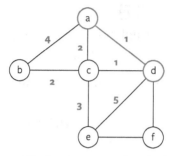

FIGURE 4.15 The first iteration of Prim's algorithm.

In the next iteration we update edges from T to all adjacent vertices b, c, e, and f. The binary heap becomes $H = \{c_1, b_4, e_5, f_7\}$. Note that the edge (a, c) to vertex c gets replaced by a shorter one (d, c). Now vertex c in H is the closest one to T. We remove it from the heap H and add it to tree T. After this step $T = \{a, d, c\}$ and $H = \{b_4, e_5, f_7\}$.

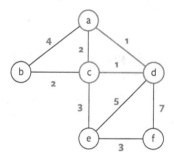

FIGURE 4.16 The second iteration of Prim's algorithm.

Next, we update edges from T to all adjacent vertices b, e, and f. The binary heap becomes $H = \{b_2, e_3, f_7\}$. Since vertex b is the shortest one, we remove it from H and add it to T. After this step $T = \{a, d, c, b\}$ and $H = \{e_5, f_7\}$.

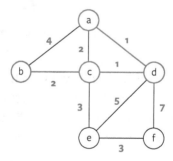

FIGURE 4.17 The third iteration of Prim's algorithm.

Again we update edges in H, so $H = \{e_3, f_7\}$. Vertex e is the shortest one; we remove it from H and add it to T. After this step $T = \{a, d, c, b, e\}$ and $H = \{f_7\}$.

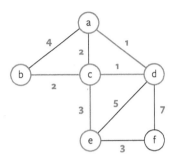

FIGURE 4.18   The fourth iteration of Prim's algorithm.

Update heap $H = \{f_3\}$. Vertex f is the shortest one; we remove it from H and add it to T. After this step $T = \{a, d, c, b, e, f\}$ and $H = \{\}$.

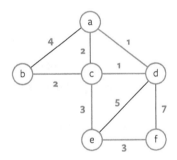

FIGURE 4.19   The minimum spanning tree.

We have constructed the minimum spanning tree of the total weight $1 + 1 + 2 + 3 + 3 = 10$.

4.4.1.2   Complexity of Prim's Algorithm

The Prim's algorithm complexity depends heavily on the chosen graph representation. The following analysis assumes using an adjacency list structure for graph representation. Here is a pseudocode for the algorithm:

```
1. H = minHeap(V);
2. insert(s, H);                     // start at vertex s
3. while (H is not empty)
4. {
5.   u = deleteMin(H);               // O(log V)
6.   for each w in adj(u)
7.   {
8.       if(weight(w,u) < key(u))
9.               key(u) = weight(w,u)  // update edge weight
10.              decreaseKey(u, H);    // O(log V)
11.  }
12. }
```

We maintain a min-heap of V vertices (line 1). In each step of the algorithm we delete a vertex (line 5) with the smallest weight (this takes $O(\log V)$ by deleteMin operation). We also update edges (lines 8–10) to all adjacent vertices (this takes $O(\log V)$ by decreaseKey operation). We run deleteMin operation once on each vertex, so the time required is $O(V \log V)$. We run decreaseKey operation once on each edge, so the time required is $O(E \log V)$. The latter requires an explanation. Consider the inner loop, lines 6–11. The number of steps in that loop depends on the degree of vertex u, which we will denote by $\deg(u)$. Thus, the complexity of the inner loop is $O(\deg(u) \log V)$. If we add the outer loop, the total runtime complexity is given by

$$\sum_{u \in V}(O(\log V) + O(\deg(u) \log V)) = V\, O(\log V) + O(\log V) \sum_{u \in V} O(\deg(u)) =$$

$$= O(V \log V) + O(\log V)\, E = O(V \log V + E \log V).$$

Prim's algorithm can be further improved by using Fibonacci heaps that provide the best runtime in theory. In this case, the algorithm complexity is $O(V \log V + E)$ amortized.

4.4.1.3 Prim's Algorithm Using an Array

This is the simplest implementation of Prim's algorithm. We use an unsorted array instead of a priority queue. Assuming the pseudocode from the previous section, let us analyze the runtime complexity. Each deleteMin (line 5) will take $O(V)$ times to find the minimum in an unsorted array. Each decreaseKey (line 10) will take $O(1)$ to update the edge weight. The total runtime complexity is given by

$$\sum_{u \in V}(O(V) + O(\deg(u))) = V\, O(V) + \sum_{u \in V} O(\deg(u)) = O(V^2) + E = O(V^2 + E).$$

4.4.1.4   Correctness of Prim's Algorithm

Given a weighted connected graph, we will prove that Prim's algorithm finds an MST. We will prove it by induction on the number of iterations. Let $S(n)$ be a spanning tree of $n < V$ vertices, constructed so far by Prim's algorithm, and a tree M of V vertices be an MST.

> *Base case*: $n = 1$. That is true, since it is just a single node and no edges.
>
> *Inductive hypothesis*: Assume $S(n)$ is a subtree of M.
>
> *Inductive step*: We need to prove that $S(n + 1)$ is also a subtree of some MST.

Let e be the edge chosen by Prim's algorithm. We need to argue that the new tree, $S(n + 1) = S(n) + \{e\}$, is a subtree of some minimum spanning tree M_1. If $e \in M$, then this is true, since by inductive hypothesis $T(n)$ is a subtree of M, thus $S(n) \cup \{e\}$ is also a subtree of M. Consider the case $e \notin M$. Adding edge e to M creates a cycle in M. Traversing the cycle, we find another edge $e^* \in M$. So, Prim's algorithm could have added e^*, but instead chose e. It follows, weight$(e) \leq$ weight(e^*), by the greedy choice property. Next, we create a new tree $M_1 = M - \{e^*\} + \{e\}$ by removing e^* from M and adding e. The total weight of M_1 is at most the weight of M. By construction, M_1 contains $S(n + 1)$. ∎

4.4.2   Kruskal's Algorithm (1956)

For any weighted undirected connected graph $G = (V, E)$, the algorithm builds a minimum spanning tree by adding edges in a sequence of non-decreasing weights. The algorithm is a bit different from Prim's algorithm; it does not maintain a single tree but instead maintains a forest (a collection of trees). Here are the algorithm steps:

1. Sort edges in non-decreasing order by weight.
2. Start with all vertices. Each vertex forms a tree.
3. Choose the minimum weight edge and join corresponding trees if it does not create a cycle. Otherwise, discard that edge.
4. Continue to merge the trees until all vertices are connected.

The proof of correctness is quite similar to the one we used for Prim's algorithm. We leave it to the reader to work out the details.

What about the runtime complexity? Sorting takes $O(E \log E)$. Then we have to check if an adding edge will cause a cycle. Using a simple graph traversal it would take $O(V)$. We must do this test for each edge in the worst case. This will take $O(E V)$ for all edges. The total runtime is $O(E \log E + E V)$. The runtime can be improved by using an advanced data structure for the cycle detection.

4.5   Shortest Path Problem

Consider a directed or undirected weighted connected graph $G = (V, E)$. One of the nodes is designated as a source s. The problem is to find the shortest directed path from s and all other vertices in the graph. By shortest path we mean a set of edges with the minimum possible sum of their weights. These shortest paths form a tree called the shortest path tree from start node s. There are many versions (and therefore algorithms) of this problem. For example, for graphs with equal-edge weights (or without edge weights) breadth-first search can be used to solve the single-source shortest path problem. In this section we consider Dijkstra's algorithm and in chapter 6.3 we will discuss the Bellman-Ford shortest path algorithm.

4.5.1   Dijkstra's Algorithm (1959)

For any positively weighted connected graph $G = (V, E)$, the algorithm finds the shortest paths between a given source and all other vertices in V. There can be many equal weight shortest paths between two vertices; the problem requires finding only one. Before going further let us develop an intuition about the algorithm. Consider the case when we know the shortest path to some vertices. Let X denote the set of such vertices. In figure 4.20 the shortest paths are indicated by labels next to the vertices.

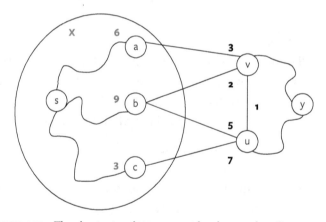

FIGURE 4.20   The shortest path to v cannot be shorter than 9.

Note that since there are no edges with negative weights, one cannot make a path shorter by visiting a vertex twice. The shortest path to vertex v consists of a path from s to a followed by the edge (a, v). This path cannot be shorter if it goes from v to u and then comes back to v, since all weights are nonnegative. By the same reason, the path s to v cannot be shorter than 9 if it does not go through vertex a. For example, the path s-c-u-y-v is of length 10 or longer.

We are now ready to define precisely Dijkstra's algorithm:

1. Start at vertex s and add it to an empty tree T. This vertex will be the root of T.
2. Expand T by adding a vertex from V\T having the minimum path length from vertex s.
3. Update distances from vertex s to adjacent vertices in V\T.
4. Continue to grow the tree until T gets all vertices, T = V.

Step 2 is a greedy choice: Among all adjacent vertices to T we pick the one that has the minimum path length from vertex s. Step 3 is the *relaxation* step: We update the path if it's shorter than in the previous instance. The greedy choice implies that we have to use an intermediate data structure, which will allow us to find a vertex with the shortest distance in the most efficient way. This suggests use of a priority queue in which we store every node v and the upper bound d(v) on its distance from the source s. Relaxing edge (u, v) means checking if we can decrease d(v) by using d(u) and the edge weight $len(u, v)$. We test whether $d(u) + len(u, v) < d(v)$. If this is true, then we found a shorter path to v, which now goes through vertex u. Thus, we update the distance to v in the priority queue.

4.5.1.1 Example

Let us run Dijkstra's algorithm on the graph in figure 4.21. We will keep a binary min-heap H as an intermediate data structure. Every element of the min-heap contains the vertex number and the path length from the source s to that vertex (marked by number next to the vertices).

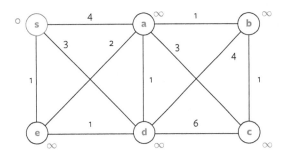

FIGURE 4.21   The Dijkstra algorithm illustrated on this graph.

We start at vertex s, so $T = \{s\}$. Edge weights from T to all adjacent vertices are $|sa| = 4, |sd| = 3, |se| = 1$. We update adjacent vertices a, d, and e in the heap $H = \{e_1, d_3, a_4\}$, where the subscripts denote a path length from s to the vertex. We pick the shortest vertex in H and add it to T. After this step $T = \{s, e\}$ and $H = \{d_3, a_4\}$.

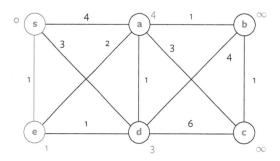

FIGURE 4.22   The first iteration of Dijkstra's algorithm.

Next, we update adjacent vertices a and d. The heap becomes $H = \{d_2, a_3\}$. The vertex with the shortest path is d. After this step $T = \{s, e, d\}$.

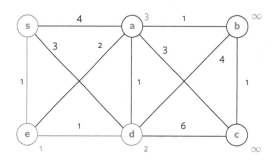

FIGURE 4.23   The second iteration of Dijkstra's algorithm.

Update adjacent vertices b and c. The heap becomes $H = \{a_3, b_6, c_8\}$. The vertex with the shortest path is a. After this step $T = \{s, e, d, a\}$.

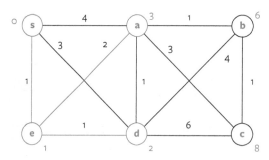

FIGURE 4.24   The third iteration of Dijkstra's algorithm.

Update the distance to b and c. The heap is $H = \{b_4, c_6\}$. The vertex with the shortest path is b. After this step $T = \{s, e, d, a, b\}$. Finally, we add c to the tree T.

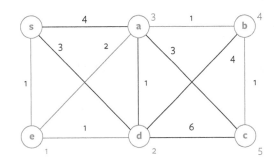

FIGURE 4.25   The tree of shortest paths from the sources.

4.5.1.2   Complexity of Dijkstra's Algorithm

Dijkstra's algorithm is quite similar to Prim's algorithm. The only difference is that Prim's algorithm stores in a priority queue a minimum cost edge whereas Dijkstra's algorithm stores the path length from a source vertex to the current vertex. If follows that the runtime of Dijkstra's algorithm using a priority queue implemented as an array is $O(V^2 + E)$, and using a min-heap is $O(V \log V + E \log V)$. The runtime complexity can be further improved by using Fibonacci heaps.

4.5.1.3 Correctness of Dijkstra's Algorithm

We will be using the following notations: $d(v)$ is the shortest s-v path, $\delta(v)$ is some s-v path that is not necessarily shortest, $\delta(v) \geq d(v)$, and $len(u, v)$ is the edge weight.

We will prove correctness of Dijkstra's algorithm by induction on the number of iterations. Let $S(n)$ denote a shortest-path tree constructed by the algorithm after n iterations.

> *Base case*: $n = 1$. That is true, since it is just a single node and no edges.
>
> *Inductive hypothesis*: Assume $S(n)$ is a shortest-path tree of n vertices.
>
> *Inductive step*: We need to prove that $S(n + 1)$ is also a shortest-path tree.

Let v be the next vertex chosen by the algorithm and let (u, v) be the chosen edge. The shortest path to vertex $u \in S(n)$ is already known; it is $d(u)$. The path to vertex v is $\delta(v) = d(u) + len(u, v)$. Assume that s-u-v path is not the shortest path. So, there is another s-v path P that is shorter. Let y be the last vertex on that path.

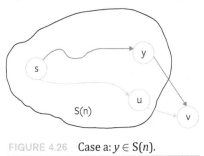

FIGURE 4.26   Case a: $y \in S(n)$.

First, note that vertex y cannot be in $S(n)$. If it was, then by relaxing edge (y, v) we would compute the distance to v as $\delta(v) = d(y) + len(y, v)$. On other hand, since P is shorter, we have $d(u) + len(u, v) > d(y) + len(y, v)$. The contradiction is $\delta(v) = d(u) + len(u, v) > d(y) + len(y, v) = \delta(v)$.

Suppose that $y \notin S(n)$. Let edge (w, x) be the first edge in P that leaves $S(n)$.

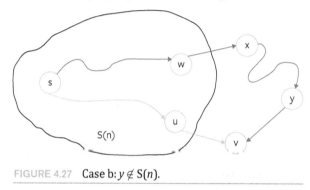

FIGURE 4.27   Case b: $y \notin S(n)$.

Vertex x may or may not be vertex y. Since x is on the real shortest path P to v that goes through vertex w, we know that $\delta(x) = d(x)$, which in turn is less than $d(v)$, since we exclude the x-y-v sub-path. Next, we note that $d(v) \le \delta(v)$, since $d(v)$ is the shortest path. Combining these together, we have

$$\delta(x) = d(x) < d(v) \le \delta(v).$$

This gives a contradiction, since the algorithm would not pick v as the next node but would instead pick x. ∎

REVIEW QUESTIONS

1. (**T/F**) In the interval scheduling problem, if all intervals are of equal size, a greedy algorithm based on earliest start time will always select the maximum number of compatible intervals.

2. (**T/F**) Any weighted undirected graph with distinct edge weights has exactly one minimum spanning tree.

3. (**T/F**) Suppose we have a graph where each edge weight value appears at most twice. Then, there are at most two minimum spanning trees in this graph.

4. (**T/F**) Kruskal's algorithm can fail in the presence of negative cost edges.

5. (**T/F**) If a connected undirected graph $G = (V, E)$ has $n = |V|$ vertices and $n + 5$ edges, we can find the minimum spanning tree of G in $O(n)$ runtime.

6. **(T/F)** The first edge added by Kruskal's algorithm can be the last edge added by Prim's algorithm.

7. **(T/F)** Suppose graph G has a unique minimum spanning tree and graph G_1 is obtained by increasing the weight of every edge in G by 1. The MST of G_1 must be different from the MST of G.

8. **(T/F)** Suppose graph G has a unique minimum spanning tree and graph G_1 is obtained by squaring the weight of every edge in G. The MST of G_1 may be different from the MST of G.

9. **(T/F)** If path P is the shortest path from u to v and w is a node on the path, then the part of path P from u to w is also the shortest path.

10. **(T/F)** If all edges in a connected undirected graph have distinct positive weights, the shortest path between any two vertices is unique.

11. **(T/F)** Suppose we have calculated the shortest paths from a source to all other vertices. If we modify the original graph G such that weights of all edges are doubled, then the shortest path tree of G is also the shortest path tree of the modified graph.

12. **(T/F)** Suppose we have calculated the shortest paths from a source to all other vertices. If we modify the original graph, G, such that weights of all edges are increased by 2, then the shortest path tree of G is also the shortest path tree of the modified graph.

EXERCISES

1. At the Perfect Programming Company, the programmers are paired in order to ensure the highest quality of produced code. The productivity of each pair is the speed of the slowest programmer. Assuming an even number of programmers, devise an efficient algorithm for pairing them up so the total productivity of all programmers is maximized.

2. A new startup, FastRoute, wants to route information along a path in a communication network, represented as a graph. Each vertex represents a router and each edge a wire between routers. The wires are weighted by the maximum bandwidth they can support. FastRoute comes to you and asks you to develop an algorithm to find the path with maximum bandwidth from any source $s_1, s_2, ..., s_k$ to any destination $t_1, t_2, ..., t_n$. Devise an algorithm that has the same runtime complexity as Dijkstra's algorithm.

3. You are given a set S of n points, labeled 1 to n, on a line. You are also given a set of k finite intervals $I_1, ..., I_k$, where each interval I_i is of the form $[s_i, e_i]$, $1 \le s_i \le e_i$. Present an efficient algorithm to find the smallest subset $X \subseteq S$ of points such that each interval contains at least one point from X. Prove that your solution is optimal.

4. You are given a minimum spanning tree T in a graph $G = (V, E)$. Suppose we remove an edge from G, creating a new graph, G_1. Assuming that G_1 is still connected, devise a linear time algorithm to find an MST in G_1.

5. You are given a minimum spanning tree T in a graph, $G = (V, E)$. Suppose we add a new edge (without introducing any new vertices) to G, creating a new graph, G_1. Devise a linear time algorithm to find an MST in G_1.

6. Given graph $G = (V, E)$ with positive edge weights, we know that Dijkstra's algorithm can be implemented in $O((E + V) \log V)$ time using a binary heap. Suppose you have been told that the input graph G is a dense graph in which $E = O(V^2)$. Find a way to implement Dijkstra's algorithm in $O(V^2)$ time.

7. Given a graph, $G = (V, E)$, whose edge weights are integers in the range $[0, W]$, where W is a relatively small integer number, we could run Dijkstra's algorithm to find the shortest distances from the start vertex to all other vertices. Design a new algorithm that will run in linear time $O(V + E)$ and therefore outperform Dijkstra's algorithm.

8. Given a directed acyclic graph, $G = (V, E)$, with nonnegative edge weights and the source s, devise a linear time algorithm to find the shortest distances from s to all other vertices.

9. You are given a graph, $G = (V, E)$, with nonnegative edge weights and the shortest path distances d(s, u) from a source vertex s to all other vertices in G. However, you are not given the shortest path tree. Devise a linear time algorithm to find a shortest path from s to a given vertex t.

10. Given a graph, $G = (V, E)$, with nonnegative edge weights and two vertices s and t, the goal is to find the shortest path from s to t with an odd number of edges. Devise an algorithm that has the same runtime complexity as Dijkstra's algorithm.

11. Given a graph, $G = (V, E)$, with nonnegative edge weights and the shortest path distances d(u, v) between any pair of vertices in G, suppose we add a new edge (without introducing any new vertices) to G, creating a new graph G_1. Devise an efficient algorithm (that outperforms Dijkstra's algorithm in the worst case) to update the shortest path distances d(u, v).

12. Given n rods of lengths $L_1, L_2, ..., L_n$, respectively, the goal is to connect all the rods to form a single rod. The length and the cost of connecting two rods are equal to the sum of their lengths. Devise an algorithm to minimize the cost of forming a single rod.

13. Given a sorted array of frequencies of size n, devise a linear time algorithm for building a Huffman tree.

Chapter 5

Divide-and-Conquer Algorithms

A DIVIDE-AND-CONQUER ALGORITHM DESIGN PARADIGM SOLVES A problem by

- dividing it into smaller subproblems of the same type;
- solving (recursively or iteratively) each subproblem; and
- combining solutions to subproblems to get solutions to the original problem.

This design approach exploits the fact that solutions to smaller subproblems used to solve larger problems. All subproblems must have exactly the same structure as the original problem and can be solved independently from each other.

Divide-and-conquer (DC) algorithms have a few advantages over other algorithmic approaches:

1. Simple proofs of correctness: The DC approach closely follows the structure of an inductive proof.
2. Efficiency: The DC approach can often lead to a more efficient solution. Its runtime complexity can be expressed by recurrences, which in most cases can be solved straightforwardly. Solving such divide-and-conquer recurrences will be a major topic of this chapter.
3. Parallelism: Independence of subproblems means that they can be solved in parallel.

As an introduction we will consider two canonical examples of DC: binary search and mergesort. Later in the chapter we will look at how to apply divide-and-conquer design technique to a variety of problems and analyze their runtime complexities.

Binary search algorithm. The algorithm finds an item in a sorted array by comparing the search item with the middle element; if they are unequal, half of the array (in which the search item cannot be) is eliminated and the search continues on to the remaining half until it is successful or that half is found to be empty. Let T(n) be the number of comparisons in the worst case needed to find an item in a sorted array of size n. We define the runtime complexity T(n) by a recurrence equation:

$$T(n) = T(n/2) + O(1)$$
$$T(1) = 1.$$

This recurrence contains the base case $T(1) = 1$ and the inductive step $T(n) = T(n/2) + O(1)$, in which we reduce a problem of size n into a subproblem of size $n/2$. On each recursive step we require a constant time $O(1)$ work to (a) find the middle element in an array and (b) compare it with the search item.

Mergesort. The algorithm sorts an array by first dividing the array into equal (or nearly equal) subarrays and then combining them in a sorted manner. Let T(n) be the number of comparisons in the worst case needed to sort an array of size n. We define the runtime complexity T(n) by the following recurrence:

$$T(n) = 2T(n/2) + O(1) + O(n)$$
$$T(1) = 1.$$

In the base case $n = 1$, the array is sorted by definition. In the inductive step we generate two subproblems of size $n/2$. We also infer the constant work of splitting the array in half and a linear time work $O(n)$ of merging two sorted arrays.

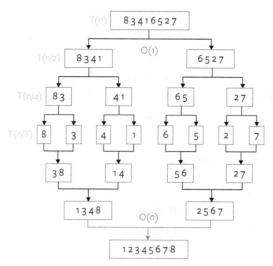

FIGURE 5.1 Mergesort example.

As we see on these two examples, divide-and-conquer algorithms follow a generic pattern: They tackle a problem of size n by recursively solving $a \geq 1$ subproblems of size n/b (where $b > 1$) and then combining the results in $f(n) > 0$ time (this also includes complexity of dividing). Therefore, the form of divide-and-conquer recurrences for the runtime complexity look like this:

$$T(n) = a\,T(n/b) + f(n)$$
$$T(1) = \Theta(1).$$

In binary search, we have $a = 1$, since we call a binary search on one half; $b = 2$, since the new subproblem size is half of the original problem size; and $f(n) = O(1)$, since finding the middle and deciding which half to recurse to takes a constant time.

In mergesort, we have $a = 2$, since we call mergesort twice; $b = 2$, since the new subproblem size is half of the original problem size; and $f(n) = O(n)$, since we merge two sorted arrays in linear time.

5.1 Solving Divide-and-Conquer Recurrences

Divide-and-conquer recurrences can be depicted as trees. The way to solve recurrences is to draw a tree of recursive calls, where each node in the tree represents a subproblem and the value at each node represents the amount of work spent at that subproblem. The root node represents the original problem. Every internal node has $a \geq 1$ children, representing the number of subproblems.

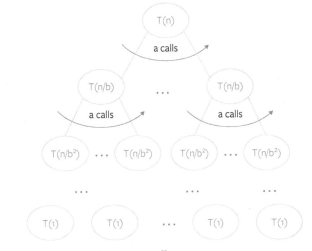

FIGURE 5.2   Tree of recursive calls.

The tree height is $h = \log_b n$, and it has $a^h = n^{\log_b a}$ leaves. This identity can be easily proven by the property of logs:

$$h = \log_b n = \frac{\log_a n}{\log_a b} = \log_a n \log_b a$$

$$a^h = (a^{\log_a n})^{\log_b a} = n^{\log_b a}.$$

To figure out how much work is being spent at each subproblem, we substitute the size of the subproblem into $f(n)$. Thus, a node for a problem size n will have a child contributing $f(n/b)$ amount of work. Note, a recurrence $T(n) = a\, T(n/b) + f(n)$ must converge, so we require $f(n/b) \leq \alpha \cdot f(n)$ for some constant $\alpha > 0$.

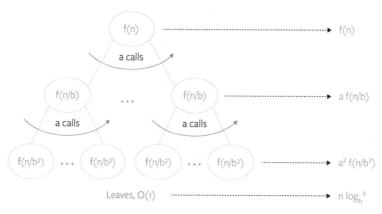

FIGURE 5.3   Tree represents the total work.

The work contributed by each leaf is constant. Once we have our tree (see figure 5.3), the total runtime $T(n)$ can be calculated by summing up the work contributed by all nodes. We can do this by summing up the work at each level of the tree and then summing up the levels of the tree. As an example, let us consider a merge-sort recursion tree:

The work at each level is n, summing up the levels lead to $T(n) = \Theta(n \log n)$ running time for mergesort.

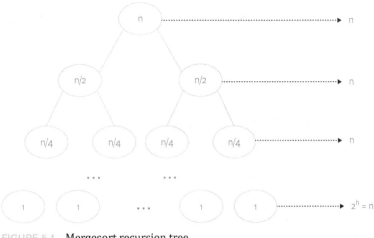

FIGURE 5.4 Mergesort recursion tree.

5.1.1 The Master Theorem

The total work depicted in figure 5.3 is given by (where the tree height is $h = \log_b n$)

$$T(n) = T(1) \cdot n^{\log_b a} + \sum_{k=0}^{h-1} a^k f\left(\frac{n}{b^k}\right)$$

and depends on three cases that may happen. Either the work done at the leaves dominates, or the work done at internal nodes dominates, or the work at all levels have about the same cost. This leads to the master theorem (here, $c = \log_b a$):

Case 1: (leaves dominate) If $f(n) = O(n^{c-\varepsilon})$, then $T(n) = \Theta(n^c)$ for some $\varepsilon > 0$.

Case 2: (all nodes) If $f(n) = \Theta(n^c \log^k n)$, $k \geq 0$, then $T(n) = \Theta(n^c \log^{k+1} n)$.

Case 3: (internal nodes dominate) If $f(n) = \Omega(n^{c+\varepsilon})$, then $T(n) = \Theta(f(n))$ for some $\varepsilon > 0$.

Let us prove Case 1: for some constant $\varepsilon > 0$

$$\text{if } f(n) = O(n^{\log_b a - \varepsilon}), \text{ then } T(n) = \Theta(n^{\log_b a}).$$

Proof. We start with simplifying the finite sum using the definition of the big-O notation:

$$\sum_{k=0}^{h-1} a^k f\left(\frac{n}{b^k}\right) \leq c \sum_{k=0}^{h-1} a^k \left(\frac{n}{b^k}\right)^{\log_b a - \varepsilon},$$

where $c > 0$ is some constant. Next, by simple algebra we get

$$c \sum_{k=0}^{h-1} a^k \left(\frac{n}{b^k}\right)^{\log_b a - \varepsilon} = c\, n^{\log_b a - \varepsilon} \sum_{k=0}^{h-1} \left(\frac{a}{b^{\log_b a}}\right)^k b^{\varepsilon\, k}.$$

Now using the properties of logs, we arrive at

$$\sum_{k=0}^{h-1} a^k f\left(\frac{n}{b^k}\right) \leq c\, n^{\log_b a}\, c \sum_{k=0}^{h-1} b^{\varepsilon\, k} = c_1\, n^{\log_b a - \varepsilon} b^{\varepsilon \log_b n} = c_1\, n^{\log_b a - \varepsilon} n^{\varepsilon} = O(n^{\log_b a}).$$

Therefore, we have showed

$$T(n) = \Theta(n^{\log_b a}) + \sum_{k=0}^{h-1} a^k f\left(\frac{n}{b^k}\right) = \Theta(n^{\log_b a}) + O(n^{\log_b a}) = \Theta(n^{\log_b a}).$$

The proof for the other two cases is left to the reader as an exercise.

5.1.2 Examples of Recurrences

Example 1. Solve the following recurrence by the master theorem:

$$T(n) = 4\,T(n/8) + n^2.$$

First, we observe that $a = 4$ and $b = 8$; next we compute $c = \log_b a = \log_8 4 = 2/3$. It follows that this is Case 3, since $f(n) = n^2 = \Omega(n^{2/3})$. Therefore, $T(n) = \Theta(n^2)$.

Example 2. Solve the following recurrence by the master theorem:

$$T(n) = 2\,T\left(\frac{n}{2}\right) + \frac{n}{\log n}.$$

We start with computing $c = \log_b a = \log_2 2 = 1$. Next, we observe that this is not Case 3, $f(n) \neq \Omega(n^{1+\varepsilon})$. We can also eliminate Case 2, $f(n) \neq \Theta(n)$, since the parameter k must be nonnegative. Finally, we claim that this does not fall into Case 1, $f(n) \neq O(n^{1-\varepsilon})$. We will prove it by contradiction. Assume that

$$f(n) = \frac{n}{\log n} = O(n^{1-\varepsilon}).$$

By the definition of Big-O notation,

$$\frac{n}{\log n} \leq c\, n^{1-\varepsilon}$$

which is the same as

$$n^{\varepsilon} \leq c \log n$$

and (after applying log to both sides)

$$\varepsilon \log n \leq \log c + \log \log n.$$

Clearly, this inequality does not hold when $n \to \infty$. It follows that the master theorem is not applicable to the original recurrence.

Example 3. Solve the following recurrence using the tree method:

$$T(n) = 2\, T\!\left(\frac{n}{2}\right) + \frac{n}{\log n}.$$

We already know that the total work (as it is depicted in figure 5.3) is given by

$$T(n) = \Theta\!\left(n^{\log_b a}\right) + \sum_{k=0}^{h-1} a^k f\!\left(\frac{n}{b^k}\right) = \Theta(n) + \sum_{k=0}^{\log(n)-1} 2^k f\!\left(\frac{n}{2^k}\right)$$

where

$$f\left(\frac{n}{2^k}\right) = \frac{n/2^k}{\log(n/2^k)}.$$

It follows,

$$T(n) = \Theta(n) + \sum_{k=0}^{\log(n)-1} 2^k \frac{n/2^k}{\log(n/2^k)} = \Theta(n) + n \sum_{k=0}^{\log(n)-1} \frac{1}{\log(n)-k}.$$

Next, we note that the finite sum can be resummed from $\log(n) - 1$ back to 0. We get

$$T(n) = \Theta(n) + n \sum_{k=0}^{\log(n)-1} \frac{1}{\log(n)-k} = \Theta(n) + n \sum_{j=1}^{\log(n)} \frac{1}{j} = \Theta(n) + n\, H_{\log n}$$

where H_k are the Harmonic numbers denoted by

$$H_n = \sum_{j=1}^{n} \frac{1}{j} = \Theta(\log n).$$

Hence,

$$T(n) = \Theta(n) + n\, H_{\log n} = \Theta(n) + n\, \Theta(\log \log n) = \Theta(n \log \log n).$$

5.2   Integer Multiplication

Given two n-digit integers a and b, our goal is to design an algorithm to compute a product $a \cdot b$. The brute force approach is to multiply two numbers digit by digit. Assuming that digit multiplication and addition are done in constant time, this leads to $\Theta(n^2)$ runtime complexity of the brute force approach.

Let us design a divide-and-conquer algorithm. We split each number in half, $a = x_1 \cdot 10^{n/2} + x_0$, $b = y_1 \cdot 10^{n/2} + y_0$, and then multiply those four pieces:

$$a \cdot b = (x_1 \cdot 10^{n/2} + x_0) \cdot (y_1 \cdot 10^{n/2} + y_0) = x_1 \cdot y_1 \cdot 10^n + (x_0 \cdot y_1 + x_1 \cdot y_0) \cdot 10^{n/2} + x_0 \cdot y_0.$$

Therefore, we reduced the problem of multiplication of two n-digit integers to multiplication of four $n/2$-digit integers. Additionally, we gained three additions, each takes

$\Theta(n)$ and two multiplications by a base, and each takes a constant time. Let $T(n)$ be a runtime complexity of multiplication of two n-digit integers, then

$$T(n) = 4\,T(n/2) + \Theta(n).$$

It follows, by the master theorem (Case 1), $T(n) = \Theta(n^2)$, which is not an improvement to the brute force approach. In 1960 A.A. Karatsuba observed that n-digit multiplication can be done with only three $n/2$-digit multiplications at the cost of increasing the number of additions:

$$a \cdot b = x_1 \cdot y_1 \cdot 10^n + ((x_0 + x_1) \cdot (y_0 + y_1) - x_0 \cdot y_0 - x_1 \cdot y_1) \cdot 10^{n/2} + x_0 \cdot y_0.$$

It looks that we have increased the number of multiplications from four to five, but that is not so, since we will compute $x_0 \cdot y_0$ and $x_1 \cdot y_1$ only once and then reuse them. The recurrence for the time complexity $T(n)$ is given now by

$$T(n) = 3\,T(n/2) + \Theta(n).$$

Using the master theorem (Case 1) we find, $T(n) = \Theta(n^{\log 3}) = \Theta(n^{1.58})$.

The eternal question is, "Can we do better?" A few years later A. Toom and S. Cook independently proposed the generalizations of the Karatsuba method by splitting an n-digit integer into three parts of size $n/3$:

$$a \cdot b = (x_2 \cdot 10^{2n/3} + x_1 \cdot 10^{n/3} + x_0) \cdot (y_2 \cdot 10^{2n/3} + y_1 \cdot 10^{n/3} + y_0).$$

However, this requires nine multiplications. To get an improvement to Karatsuba's algorithm, the number multiplication must be reduced to five. It turns out it is possible to define five new variables z_k to express x_k and y_k in terms of z_k as follows:

$$x_0 \cdot y_0 = z_0$$

$$12\,(x_1 \cdot y_0 + x_0 \cdot y_1) = 8z_1 - z_2 - 8z_3 + z_4$$

$$24\,(x_2 \cdot y_0 + x_1 \cdot y_1 + x_0 \cdot y_2) = -30z_0 + 16z_1 - z_2 + 16z_3 - z_4$$

$$12\,(x_2 \cdot y_1 + x_1 \cdot y_2) = -2z_1 + z_2 + 2z_3 - z_4$$

$$24\,x_2 \cdot y_2 = 6z_0 - 4z_1 + z_2 - 4z_3 + z_4.$$

The recurrence for the time complexity T(n) now is

$$T(n) = 5\,T(n/3) + \Theta(n)$$

and its solution is $T(n) = \Theta(n^{\log_3 5}) = \Theta(n^{1.47})$. A. Toom and S. Cook have still further generalized this idea by proposing k-way splitting, and they were able to reduce the number of multiplications from k^2 to $2k - 1$. This leads to the following recurrence

$$T(n) = (2k - 1)\,T(n/k) + \Theta(n)$$

and its solution, $T(n) = \Theta(n^{\log_k (2k-1)})$. It should be noted that by increasing k we get faster and faster algorithms; however, we will never get a linear performance. Also, we have to mention that the cost of the extra additions is growing very rapidly.

5.3   Matrix Multiplication

Given two $n \times n$ matrices, A and B, our goal is to design an algorithm to compute a product $C = A \cdot B$. The standard matrix multiplication algorithm is based on the mathematical definition of matrix multiplication in which rows of one matrix are multiplied by the column of another matrix:

$$\left(\begin{array}{cc} a_{11} & a_{12} \\ a_{21} & a_{22} \end{array} \right) \left(\begin{array}{cc} b_{11} & b_{12} \\ b_{21} & b_{22} \end{array} \right) = \left(\begin{array}{cc} a_{11}b_{11} + a_{12}b_{21} & a_{11}b_{12} + a_{12}b_{22} \\ a_{21}b_{11} + a_{22}b_{21} & a_{21}b_{12} + a_{22}b_{22} \end{array} \right)$$

For $n \times n$ matrices the runtime is $\Theta(n^3)$. In 1969 V. Strassen, inspired by Karatsuba's method, designed a divide-and-conquer algorithm for matrix multiplication. The idea is to divide the original matrix in four matrices, each of which of size $n/2 \times n/2$, and then multiply them using the definition of block-matrix multiplication. Let us assume that n is a power of two and write matrices A and B as block matrices:

$$A = \left(\begin{array}{cc} A_{11} & A_{12} \\ A_{21} & A_{22} \end{array} \right), B = \left(\begin{array}{cc} B_{11} & B_{12} \\ B_{21} & B_{22} \end{array} \right), C = AB = \left(\begin{array}{cc} C_{11} & C_{12} \\ C_{21} & C_{22} \end{array} \right).$$

The usual matrix multiplication works by substituting the blocks into the formula. Each of the four block entries of C are computed independently from one another; thus, we may come up with the following recurrence for the runtime complexity:

$$T(n) = 8\,T(n/2) + \Theta(n^2).$$

On each step we compute four block matrices, C_{ij}, each requiring two recursive calls to matrices of size $n/2 \times n/2$. Additionally, the algorithm requires four matrix additions, each taking $\Theta(n^2)$. Using the master theorem (Case 1) we find $T(n) = \Theta(n^3)$. This is not an improvement to the standard matrix multiplication. V. Strassen has observed that the number of block matrix multiplications can be reduced to seven by defining new matrices:

$$S_1 = (A_{12} - A_{22})(B_{21} + B_{22})$$

$$S_2 = (A_{11} + A_{22})(B_{11} + B_{22})$$

$$S_3 = (A_{11} - A_{21})(B_{11} + B_{12})$$

$$S_4 = (A_{11} + A_{12}) B_{22}$$

$$S_5 = A_{11}(B_{12} - B_{22})$$

$$S_6 = A_{22}(B_{21} - B_{11})$$

$$S_7 = (A_{21} + A_{22}) B_{11}$$

so that

$$
\begin{pmatrix} A_{11} & A_{12} \\ A_{21} & A_{22} \end{pmatrix}
\begin{pmatrix} B_{11} & B_{12} \\ B_{21} & B_{22} \end{pmatrix}
=
\begin{pmatrix} S_1 + S_2 - S_4 + S_6 & S_4 - S_5 \\ S_6 + S_7 & S_2 - S_3 + S_5 - S_7 \end{pmatrix}.
$$

The recurrence for the time complexity $T(n)$ now is

$$T(n) = 7\,T(n/2) + \Theta(n^2)$$

and its solution is $T(n) = \Theta(n^{\log 7}) = \Theta(n^{2.808})$. We got a faster algorithm at the cost of increasing the number of additions, Strassen's algorithm requires 18 matrix additions. The algorithm also requires significantly more memory compared to the standard algorithm.

There are more recent algorithms that are theoretically faster than Strassen:

1969, Strassen $O(n^{2.808})$
1978, Pan $O(n^{2.796})$

1979, Bini $O(n^{2.78})$

1981, Schonhage $O(n^{2.548})$

1981, Pan $O(n^{2.522})$

1982, Romani $O(n^{2.517})$

1982, Coppersmith and Winograd $O(n^{2.496})$

1986, Strassen $O(n^{2.479})$

1989, Coppersmith and Winograd $O(n^{2.376})$

2010, Stothers $O(n^{2.374})$

2011, Williams $O(n^{2.3728642})$

2014, Le Gall $O(n^{2.3728639})$

However, the constant factor hidden in the upper bounds is so large that these algorithms are only valuable for matrices of enormous sizes. Even Strassen's algorithm is not beneficial on current architectures for matrix sizes below 500.

5.4   The Maximum Subsequence Sum Problem

Given an array A of n numbers, design a divide-and-conquer algorithm that finds a subarray such that $A[i] + A[i + 1] + ... + A[j]$ is the maximum. For example, $A = \{3, -4, 5, -2, -2, 6, -3, 5, -3, 2\}$. The maximum sum subarray is $\{5, -2, -2, 6, -3, 5\}$.

The problem is easy when all the numbers are positive (then the entire array is the maximum) or negative (then we need to find the maximum number in the array). The problem becomes interesting when the array contains positive and negative numbers. Let's start with the brute force algorithm: we generate all subarrays (there are $\Theta(n^2)$ of them) and then find the one with the maximum sum. This is a cubic time $\Theta(n^3)$ algorithm.

The divide-and-conquer approach involves splitting the array in half by the median index and making recursive calls on each half. This will find the maximum subarray in the left half and the maximum subarray in the right half. But the solution to the problem may not necessarily be included entirely within the left or right subarrays. It may span both subarrays. Therefore, in the combining step span(n) we need to search for the maximum subarray that begins in the left half of the array and ends in the right half. An overall maximum is then returned as the maximum of the three (left, right, and span). Let $T(n)$ be a runtime complexity of finding the subarray of maximum sum. Then,

$$T(n) = 2\,T(n/2) + \text{span}(n).$$

The combine step span(n) requires a linear search from the middle index of A to the left and to the right. Therefore,

$$T(n) = 2\,T(n/2) + \Theta(n).$$

Solving the above recurrence by the master theorem yields $T(n) = \Theta(n \log n)$.

Let us briefly explain the implementation of span(n) in linear time based on the above example. If the solution spans the center, then it must include the middle elements -2 and the next to it 6:

$$\{3, -4, 5, -2, -2, 6, -3, 5, -3, 2\}.$$

Next, we start with -2 and go left computing partial sums:

$$\{0, -3, 1, -4, -2, 6, -3, 5, -3, 2\}.$$

Then we compute partial sums to the right starting with 6:

$$\{0, -3, 1, -4, -2, 6, 3, 8, 5, 7\}.$$

We choose the max value from each side: $1 + 8 = 9$

$$\{0, -3, 1, -4, -2, 6, 3, 8, 5, 7\}.$$

It follows that the maximum sum subarray is $\{5, -2, -2, 6, -3, 5\}$.

In conclusion, we have to mention that there are faster algorithms for solving this problem; however, they do not use a divide-and-conquer technique.

5.5   Computing Fibonacci Numbers

The Fibonacci numbers are defined by the recurrence relation $F_n = F_{n-1} + F_{n-2}, n \geq 2$ with the base values $F_0 = 0$ and $F_1 = 1$. The formal definition of this sequence directly maps to a divide-and-conquer algorithm to compute the n-th Fibonacci number F_n. Here is a pseudocode for the algorithm:

```
int fib(int n) {
   if (n == 0 || n == 1) return 1
   else
   return fib(n-1) + fib(n-2)
}
```

Its runtime complexity T(n) can be expressed as

$$T(n) = T(n-1) + T(n-2) + \Theta(1)$$

assuming that two Fibonacci numbers can be added in constant time. The solution to the recurrence is exponential in n; we roughly double the work on each recursive call. For large n the addition of Fibonacci numbers F_n takes a linear time in the number of bits. The reason is that $F_n = \Theta(\varphi^n)$, where φ is a golden ratio, and

$$\log(F_n) = \log(\Theta(\varphi^n)) = \Theta(n\log(\varphi)) = \Theta(n).$$

Therefore, the algorithm runtime complexity with non-constant time arithmetic is given by

$$T(n) = T(n-1) + T(n-2) + \Theta(n).$$

Its solution is also exponential in n.

For this problem, divide and conquer ends up having exponential runtime complexity just because the recurrence tree for T(n) has a height $\Theta(n)$ and an exponential number of nodes.

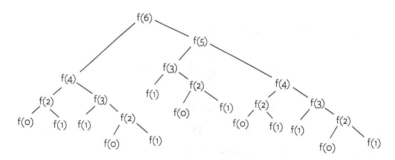

FIGURE 5.5   A recurrence tree for F_6.

However, it turns out that only some of these nodes are distinct, the rest are repeats. Figure 5.5 demonstrates redundant computations for F_6; we recompute the same Fibonacci numbers over and over again.

One may wonder why a divide-and-conquer approach was so efficient for merge-sort. The reason is that a recurrence tree for mergesort (see figure 5.4) has a height $\Theta(\log n)$ and therefore a polynomial number of nodes. As the result of this, we shall use a divide-and-conquer technique only when subproblems are independent. In case of overlapping subproblems, the better time complexity may be obtained by a dynamic programming approach.

1. (**T/F**) For a divide-and-conquer algorithm, it is possible that the dividing step takes asymptotically longer time than the combining step.

2. (**T/F**) A divide-and-conquer algorithm acting on an input size of n can have a lower bound less than $\Theta(n \log n)$.

3. (**T/F**) There exist some problems that can be efficiently solved by a divide-and-conquer algorithm but cannot be solved by a greedy algorithm.

4. (**T/F**) It is possible for a divide-and-conquer algorithm to have an exponential runtime.

5. (**T/F**) A divide-and-conquer algorithm is always recursive.

6. (**T/F**) The master theorem can be applied to the following recurrence: $T(n) = 1.2\, T(n/2) + n$.

7. (**T/F**) The master theorem can be applied to the following recurrence: $T(n) = 9\, T(n/3) - n^2 \log n + n$.

8. (**T/F**) Karatsuba's algorithm reduces the number of multiplications from four to three.

9. (**T/F**) The runtime complexity of mergesort can be asymptotically improved by recursively splitting an array into three parts (rather than into two parts).

10. (**T/F**) Two $n \times n$ matrices of integers are multiplied in $\Theta(n^2)$ time.

11. (**Fill in the blank**) Let A, B be two 2×2 matrices that are multiplied using the standard multiplication method and Strassen's method.

    a. Number of multiplications in the standard method: _____

    b. Number of additions in the standard method: _____

    c. Number of multiplications using Strassen's method: _____

    d. Number of additions using Strassen's method: _____

12. (**Fill in the blank**) The space complexity of Strassen's algorithm is: _____.

EXERCISES

1. Solve

$$T(n) = 3\, T(n/4) + n$$

by the recurrence tree method.

2. Solve

$$T(n) = T(3n/4) + T(n/4) + n$$

by the recurrence tree method.

3. Solve the following recurrences by the master theorem:

$$T(n) = 3\,T\left(\frac{n}{2}\right) + n\log n$$

$$T(n) = 8\,T\left(\frac{n}{6}\right) + \log n$$

$$T(n) = 16\,T\left(\frac{n}{4}\right) + \frac{n}{\log n}$$

$$T(n) = \sqrt{7}\,T\left(\frac{n}{2}\right) + n^{\sqrt{3}}$$

$$T(n) = 10\,T\left(\frac{n}{2}\right) + 2^{n}$$

4. Prove Case 2 of the Master theorem.

5. Prove Case 3 of the Master theorem.

6. There are two sorted arrays, each of size n. Design a divide-and-conquer algorithm to find the median of the array obtained after merging the 2 arrays. Discuss its worst-case runtime complexity.

7. You are given an unsorted array of all integers in the range $[0, ..., 2^k - 1]$ except for one integer, which is denoted by M. Describe a divide-and-conquer algorithm to find the missing number M and discuss its worst-case runtime complexity in terms of $n = 2^k$.

8. We know that binary search on a sorted array of size n takes $\Theta(\log n)$ time. Design a similar divide-and-conquer algorithm for searching in a sorted *singly linked list* of size n. Discuss its worst-case runtime complexity.

9. We know that mergesort takes $\Theta(n \log n)$ time to sort an array of items. Design a divide-and-conquer mergesort algorithm for sorting a singly linked list. Discuss its worst-case runtime complexity.

10. Given a sorted array of n integers that has been rotated an unknown number of times, give an $\Theta(\log n)$ divide-and-conquer algorithm that finds an element in the resulting array. Note, after a single rotation, the array is not sorted anymore, so we cannot use the binary search. An example of a rotations sorted array is $A = [1, 3, 5, 7, 11]$; after first rotation it is $A = [3, 5, 7, 11, 1]$, and after second rotation it is $A = [5, 7, 11, 1, 3]$. You may assume that that array has no duplicates.

11. Consider a two-dimensional array A of size $n \times n$ filled with integers. In the array each row is sorted in ascending order and each column is also sorted in ascending order. Our goal is to determine if a given value x exists in the array. Design a divide-and-conquer algorithm to solve this problem and state the runtime of your algorithm. Don't just call binary search on each row or column. Your algorithm should take strictly less than $O(n^2)$ time to run.

12. Improve your divide-and-conquer algorithm from Exercise 10 to run in $\Theta(n)$ time.

13. A polygon is called convex if all its internal angles are less than 180°. A convex polygon is represented as an array V with n vertices of the polygon, where each vertex is in the form of a coordinate pair (x, y). We are told that $V[1]$ is the vertex with the least x coordinate and that the vertices $V[1], V[2], ..., V[n]$ are ordered counter-clockwise. Design a divide-and-conquer algorithm to find the vertex with the largest x-coordinate. Discuss its worst-case runtime complexity.

Chapter 6

Dynamic Programming

I N THIS CHAPTER WE WILL LEARN another powerful algorithm design technique that is used to solve a broad variety of problems by breaking them down into simpler subproblems and storing their solutions for further computation. We usually apply dynamic programming to optimization problems.

The technique of dynamic programming (usually referred to as DP) was originally introduced by Richard Bellman in the 1950s. At that time there was no programming as we understand it today; the word *computer* meant a person performing mathematical calculations. In that time early computers were mostly women who used painstaking calculations on paper and later on punch cards. The Turing machine that describes a model for algorithms and computational problem solving was widely adapted only in the 1960s. Originally R. Bellman referred the word *programming* to the use of the method to find an optimal *program*, in the sense of planning or scheduling. The word *dynamic* was chosen by R. Bellman to capture the multistage solution to a problem.

6.1   Introduction

There are *two key attributes* that a problem must have in order for dynamic programming to be applicable:

- *Optimal substructure*: The solution can be obtained by the combination of optimal solutions to its subproblems. Such optimal substructures are usually described recursively.

- *Overlapping subproblems:* The space of subproblems must be small, so an algorithm solving the problem should solve the same subproblems over and over again.

Reading this you may be wondering how dynamic programming differs from a greedy approach. The major difference is that greedy algorithms first make a greedy choice and then solve the resulting subproblems. Dynamic programming is similar to brute force and will examine all subproblems. A DP algorithm can be described as a multi-stage decision process, and therefore we can construct a recurrence tree to enumerate all possible subproblems. During the DP algorithm execution, we have to consider all available choices at any given node. In the greedy model we use a greedy heuristic to pick just one choice.

Comparing DP to a divide-and-conquer algorithm, we say that dynamic programming usually enumerates all possible dividing strategies and therefore extends divide and conquer by reusing subproblems solutions. Divide-and-conquer partitions the problem into disjointed subproblems, though dynamic programming applies when the subproblems overlap. We may view a divide-and-conquer algorithm as a DP with no subproblem overlapping. The efficiency of DP directly depends on the amount of subproblem overlapping; the more overlapping we have, the more efficient DP algorithm we get.

A dynamic programming algorithm is implemented either recursively (*memoization*) or iteratively (*tabulation*) by placing all intermediate results into a table. Let us explain the differences between the two techniques on the example of Fibonacci numbers F_n. In Chapter 5.5 we demonstrated a divide-and-conquer approach to computing the Fibonacci numbers. We have shown that divide and conquer ends up having exponential time complexity, mainly due to the exponential number of overlapping subproblems. One way to avoid redundant computation is *memoization*. Memoization is a recursive optimization technique to speed up recursive programs by storing the intermediate results in a table. Here is a pseudocode using memoization:

```
int table [50];   //initialize to zero
table[0] = table[1] = 1;
int fib(int n) {
  if (table[n] == 0)
      table[n] = fib(n-1) + fib(n-2);
  return table[n];
}
```

The runtime complexity T(n) of this implementation is given by

$$T(n) = T(n-1) + O(n).$$

Note that the complexity of fib (n–2) is constant since that Fibonacci number will be computed during a call to fib (n–1). The solution to this recurrence is $\Theta(n^2)$. This example demonstrates that reusing previously computed (and stored) values leads to a more efficient algorithm. Next, we consider *tabulation*: a non-recursive bottom-up optimization technique. Here is a pseudocode using tabulation:

```
int table [50];
table[0] = table[1] = 1;
int fib(int n) {
   for(int k = 2; k < n; k⁺⁺)
       table[k] = table[k–1] + table[k–2];
}
```

It has the same runtime complexity as memoization. Generally, dynamic programming techniques can be implemented either using tabulation (a non-recursive bottom-up approach) or memoization (a recursive top-down approach). Both results in the same solution, though they may differ by a constant factor in runtime and memory use. For all DP algorithms in this chapter we will always use tabulation as the implementation approach.

In conclusion of the introduction we note that the example of Fibonacci numbers demonstrates that reusing previously computed (and stored) subproblems *may* lead to a more efficient algorithm. The important aspect is that the total number of unique subproblems to be solved must be polynomial.

6.2   Knapsack Problem

You are given a set of n unique items, with weights w_1, ..., w_n and values v_1, ..., v_n, where the weights and values are all integers. The problem is to find a subset of the most valuable items such that their total weight does not exceed W. We assume that all items are unbreakable (thus, 0-1 problem).

Let's start with the brute force algorithm: Consider all possible subsets of n items and then find the one with the maximum value. The worst-case runtime complexity of this approach is exponential, since the total number of subsets is $O(2^n)$; there are two choices for each item: Either we pick that item, or we don't.

Next, we turn to dynamic programming by storing all distinct subproblems (subsets) in a table. First, we formalize the problem by introducing an indicator variable x_k for each item $k = 1, 2, ..., n$:

$$x_k = \begin{cases} 1, \text{ if item } k \text{ is selected} \\ 0, \text{ otherwise} \end{cases}.$$

Then, we write the 0-1 Knapsack problem as follows

$$\max \sum_{k=1}^{n} v_k x_k$$

$$\sum_{k=1}^{n} w_k x_k \leq W$$

This formalization helps us to visualize decisions we make, which in turn will help us to define subproblems. Figure 6.1 shows that we start with n items and an empty knapsack of capacity W. The first decision we make is to either select the nth item (the left child in the tree) or not select it (the right child in the tree)

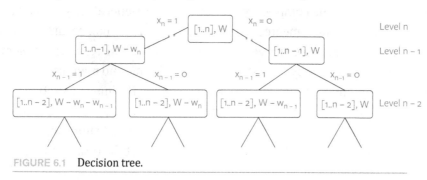

FIGURE 6.1   Decision tree.

If an item is selected, the knapsack capacity gets smaller; also, the set of available items shrinks by one (remember, all items are unique). Each node in this tree represents a subproblem, call it OPT[k, w], that corresponds to the maximum value achievable using a knapsack of capacity $0 \leq w \leq W$ and items 1, 2, ..., k, where $1 \leq k \leq n$. In order to compute OPT[k, w] we need to express it in terms of the smaller subproblems. Again, this tree suggests two cases:

1. $x_k = 1$, k-th item is included

    OPT[k, w] $= v_k +$ OPT[$k - 1$, $w - w_k$]

2. $x_k = 0$, k-th item is not included

    OPT[k, w] $=$ OPT[$k - 1$, w]

We do not know if the k-th item is actually included or not into the optimal solution; therefore, we have to try both possibilities and then choose the maximum:

$$\text{OPT}[k,w] = \max(v_k + \text{OPT}[k-1, w - w_k], \text{OPT}[k-1, w]).$$

The optimal solution we seek is OPT[n, W]. This recursive definition must be terminated by base cases:

$$OPT[k, w] = 0, \text{ if } k = 0 \text{ or } w = 0$$
$$OPT[k, w] = OPT[k - 1, w], \text{ if } w_k > w.$$

The first base case represents a situation when the knapsack has zero capacity or there are no items to choose from. The second base case occurs when the item to choose is too big for the knapsack (remember, items are not breakable). The algorithm then consists of filling out a two-dimensional table. We fill out a table in the bottom-up manner, from smaller size subproblems to larger ones.

Let us trace the algorithm on the following example: $n = 4$, $W = 5$ and $(w_k, v_k) = \{(2, 3), (3, 4), (5, 5), (5,6)\}$. Let OPT[$k$, w] be a table (see table 6.1) where each row represents available items $k = 0, 1, 2, 3, 4$, and each column represents the knapsack capacities in the weight units.

TABLE 6.1   OPT[k, w] table filled with initial conditions

	0	1	2	3	4	5
0	0	0	0	0	0	0
1	0					
12	0					
123	0					
1234	0					

The recursive definition of OPT[k, w] infers that to enter a value at a given (k, w) index; we have to know table entries at $(k - 1, w)$ and $(k - 1, w - w_k)$. This suggests filling up the table from top to bottom and from left to right.

Table 6.2 demonstrates the case when only the first item $(w_1, v_1) = (2, 3)$ is available. To enter, for example, OPT[1, 2], we need to lookup OPT[0, 2] (the first item is not chosen) and OPT[0, 0] (the first item is chosen). Thus, OPT[1, 2] = max(3 + 0, 0) = 3.

TABLE 6.2   OPT[k, w] for the first item (w_1, v_1)

	0	1	2	3	4	5
0	0	0	0	0	0	0
1	0	0	3	3	3	3
12	0					
123	0					
1234	0					

Table 6.3 demonstrates the case when two items, $(w_1, v_1) = (2, 3)$ and $(w_2, v_2) = (3, 4)$, are available. In order to calculate, for example, OPT[2, 5], we need to lookup OPT[1, 5] and OPT[1, 2]. Thus, OPT[2, 5] = max(4 + 3, 3) = 7.

TABLE 6.3   OPT[k, w] for the first two items

	0	1	2	3	4	5
0	O	O	O	O	O	O
1	O	O	3	3	3	3
12	O	O	3	4	4	7
123	O					
1234	O					

The final OPT[k, w] is shown in table 6.4. The optimal solution is OPT[4, 5] = 7, that means that we found a subset of items with the maximum value 7.

TABLE 6.4   The final OPT[k, w] table

	0	1	2	3	4	5
0	O	O	O	O	O	O
1	O	O	3	3	3	3
12	O	O	3	4	4	7
123	O	O	3	4	5	7
1234	O	O	3	4	5	7

Here is a pseudocode. We fill a two-dimensional table with $n + 1$ rows and $W + 1$ columns:

```
int knapsack(int W, int w[ ], int v[ ], int n) {
   int OPT [n+1][ W+1];
   for (k = 0; k <= n; k⁺⁺) {
      for (j = 0; j <= W; j⁺⁺) {
         if (k==0 || j==0) OPT [k][j] = 0;
         if (w[k] > j) OPT [k][j] = OPT [k-1][j];
         else
            OPT [k][j] = max(v[k] + OPT [k-1][j - w[k-1]],  OPT [k-1][j]);
      }
   }
   return Opt[n][W];
}
```

Each table entry takes constant time to fill, since the work we do involves two table lookups and one comparison. The overall running time is $O(n\,W)$.

Note that the OPT[k, w] table does not show the optimal items, but only the maximum value. We can trace back in the table (see table 6.5) to find which items give us that value. Starting from OPT[n, W], we check if OPT[n, W] = OPT[$n - 1, W$]. If they are equal, it means the nth item was not chosen, then go to OPT[$n - 1, W$]. If they are not equal, return the nth item and go to OPT[$n - 1, W - w_n$]. Continue until you reach one of the base cases.

Table 6.5 illustrates that by tracing back we find the optimal solution consisting of two items $(w_1, v_1) = (2, 3)$ and $(w_2, v_2) = (3, 4)$.

TABLE 6.5 Tracing OPT[k, w]

	0	1	2	3	4	5
0	O	O	O	O	O	O
1	O	O	3	3	3	3
12	O	O	3	4	4	7
123	O	O	3	4	5	7
1234	O	O	3	4	5	7

6.2.1 Pseudo-Polynomial Running Time

The solution to the knapsack problem is not polynomial in the input size, but *pseudo-polynomial*. This section explains the subtle difference between the two.

Let us compute the total input size of the knapsack problem:

```
int knapsack(int W, int w[], int v[], int n).
```

An array of weights `int w[ ]` will take $O(\log w_1) + \dots + O(\log w_n)$ bits. Assuming that each item does not exceed the knapsack capacity W, this simplifies to $O(n \log W)$. An array of values `int v[ ]` will take $O(\log v_1) + \dots + O(\log v_n) = O(n \log V)$ bits, where $V = \max(v_1, \dots, v_n)$. Thus, the total input size is

$$O(\log W + n \log W + n \log V + \log n) = O(n \log (W V))$$

bits. Now compare the input size with the running time $O(n\,W)$. Is $O(n \cdot W)$ polynomial in the input size $O(n \cdot \log W)$? It is polynomial in n, but it is not polynomial in W. Let $k = \log W$; then the input size is $O(n \cdot k)$ and the running time is $O(n \cdot 2^k)$. This is an exponential time algorithm! Indeed, if we increase the knapsack capacity by 2, so $k' = 2 \cdot k$, the running time $O(n \cdot 2^{k'}) = O(n \cdot 2^{2k})$ will increase quadratically.

Definition. A numeric algorithm runs in pseudo-polynomial time if its running time is polynomial in the numeric value of the input but is exponential in the input size.

We will see in Chapter 9 that it is not known if the knapsack problem can be solved in polynomial time. It is also not proven that it cannot be solved in polynomial time.

6.3 Static Optimal Binary Search Tree

In this section we will solve the optimization problem of finding the binary search tree that minimizes the total search time, given a set of keys and probabilities of looking up each key. The tree cannot be modified (no insertions and deletions) after it has been constructed.

We are given a sequence $k_1 < k_2 < ... < k_n$ of n keys, which are to be stored in a binary search tree. We are also given a search probability p_i for each key k_i. The search cost for key k_i is defined by $depth(k_i)$, where we assume that the root depth is 1 (for convenience of computations). We need to build a binary search tree T from the keys with the minimum total search cost:

$$Cost(T) = \sum_{i=1}^{n} p_i \cdot depth_T(k_i)$$

Example. Consider 5 keys $k_1 < k_2 < k_3 < k_4 < k_5$ with the following search probabilities: $p_1 = 0.25$, $p_2 = 0.2$, $p_3 = 0.1$, $p_4 = 0.15$, and $p_5 = 0.3$. There are many different binary search trees where the given keys can be stored. Here is one possibility (a balanced tree) with the total cost 2.2:

$$Cost = (0.25 \times 2) + (0.2 \times 1) + (0.1 \times 3) + (0.15 \times 2) + (0.3 \times 3) = 2.2.$$

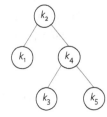

k_i	Depth (k_i)	$p_i \bullet$ depth (k_i)
1	2	0.5
2	1	0.2
3	3	0.3
4	2	0.3
5	3	0.9

FIGURE 6.2   A balanced tree.

There is another possibility—a greedy tree. The greedy approach inserts the most frequent key first.

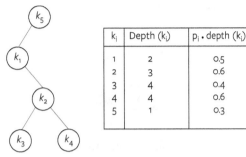

k_i	Depth (k_i)	$p_i \cdot$ depth (k_i)
1	2	0.5
2	3	0.6
3	4	0.4
4	4	0.6
5	1	0.3

FIGURE 6.3   A greedy tree.

The total cost of the greedy tree is 2.4. Finally is an optimal tree of the cost of 2.15.

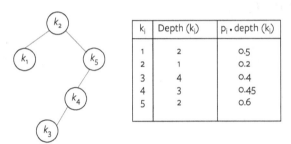

k_i	Depth (k_i)	$p_i \cdot$ depth (k_i)
1	2	0.5
2	1	0.2
3	4	0.4
4	3	0.45
5	2	0.6

FIGURE 6.4   An optimal tree.

This example demonstrates that an optimal BST may not have the smallest height nor have the highest probability key at the root. Also, an optimal BST is different from the Huffman tree (chapter 4.3), since the keys are not restricted to be leaves only. This suggests a brute force approach when we consider all possible binary trees and then choose the optimal. The only problem is that there are exponentially many binary trees (they are counted in the Catalan numbers[1]). Fortunately, by using dynamic programming, we can solve the problem efficiently.

The idea behind the DP approach is that in order to find an optimal solution for all keys k_1, ..., k_n, we must be able to find an optimal solution for any subset k_i, ..., k_j. Let OPT[i, j] be the total search cost for the optimal tree T on k_i, ..., k_j keys.

$$OPT[i, j] = \sum_{s=i}^{j} p_s \cdot depth_T(k_s)$$

1   "Catalan numbers," Wikipedia, https://en.wikipedia.org/wiki/Catalan_number

For this to work, we need to express OPT[i, j] in terms of smaller subproblems. Suppose the root of T on $k_i, ..., k_j$ keys is k_r, where $i \leq r \leq j$. This breaks the tree T into to subtrees: T_L — a subtree on $k_i, ..., k_{r-1}$ keys, and T_R — a subtree on $k_{r+1}, ..., k_j$ keys.

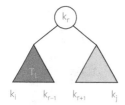

FIGURE 6.5   Computing subproblems.

We can therefore compute OPT[i, j] as follows:

$$\text{OPT}[i, j] = \sum_{s=i}^{r-1} p_s \cdot depth_T(k_s) + p_r + \sum_{s=r+1}^{j} p_s \cdot depth_T(k_s).$$

Note that $depth_T(k_s) = 1 + depth_{T_L}(k_s)$ and $depth_T(k_s) = 1 + depth_{T_R}(k_s)$. It follows that

$$\text{OPT}[i, j] = \sum_{s=i}^{r-1} p_s \cdot (1 + depth_T(k_s)) + p_r + \sum_{s=r+1}^{j} p_s \cdot (1 + depth_T(k_s))$$

$$= p_i + ... + p_j + \sum_{s=i}^{r-1} p_s \cdot depth_{T_L}(k_s) + \sum_{s=r+1}^{j} p_s \cdot depth_{T_R}(k_s)$$

$$= p_i + ... + p_j + \text{OPT}[i, r-1] + \text{OPT}[r+1, j].$$

Finally, since we don't know the r, we minimize OPT[i, j] over all choices of r, giving us the final recurrence

$$\text{OPT}[i, j] = \min_{i \leq r \leq j} \{ p_i + ... + p_j + \text{OPT}[i, r-1] + \text{OPT}[r+1, j] \}$$

with two base cases

$$\text{OPT}[i, i] = p_i$$
$$\text{OPT}[i, i-1] = 0.$$

The optimal solution we seek is OPT[$1, n$].

Runtime. There are n^2 subproblems, and each subproblem takes $O(n)$ time to compute. Thus, the total running time is $O(n^3)$.

Let us trace the algorithm on keys $k_1 < k_2 < k_3 < k_4 < k_5$ with the following search probabilities: $p_1 = 0.25$, $p_2 = 0.2$, $p_3 = 0.1$, $p_4 = 0.15$, and $p_5 = 0.3$. We will assume that keys are just numbers 1, 2, 3, 4, and 5. Table 6.6 shows the OPT[i, j] table filled with initial conditions.

TABLE 6.6 OPT[i, j] table filled with initial conditions.

	0	1	2	3	4	5
1	0	0.25				
2		0	0.2			
3			0	0.1		
4				0	0.15	
5					0	0.3

The recursive definition of OPT[i, j] infers that we fill up the table diagonally. The first value to compute is OPT[1, 2]

$$OPT[1, 2] = p_1 + p_2 + \min_{1 \leq r \leq 2}\{OPT[1, r-1] + OPT[r+1, 2]\}$$

$$OPT[1, 2] = 0.45 + \min(0 + 0.2, 0.25 + 0) = 0.65.$$

Proceeding in the same way, we fill up the whole diagonal OPT[$i, i + 1$], as shown in table 6.7.

TABLE 6.7 Diagonal OPT[$i, i+1$] for $i = 1, 2, 3, 4$

	0	1	2	3	4	5
1	0	0.25	0.65			
2		0	0.2	0.4		
3			0	0.1	0.35	
4				0	0.15	0.6
5					0	0.3

Next, we fill up the next diagonal OPT[$i, i+2$] for $i = 1, 2, 3$. To compute, for example, OPT[1, 3] we do the following

$$OPT[1, 3] = p_1 + p_2 + p_3 + \min_{1 \leq r \leq 3}\{OPT[1, r-1] + OPT[r+1, 3]\}$$

$$OPT[1, 3] = 0.55 + \min(0 + 0.4, 0.25 + 0.1, 0.65 + 0) = 0.9.$$

Tables 6.8 shows the final result.

TABLE 6.8 The final cost table

	0	1	2	3	4	5
1	0	0.25	0.65	0.9	1.3	2.15
2		0	0.2	0.4	0.8	1.45
3			0	0.1	0.35	0.9
4				0	0.15	0.6
5					0	0.3

In order to compute the actual BST, for each subproblem we need also to store the root of the corresponding subtree

$$\text{root}[i,\,j] = \operatorname*{argmin}_{i \le r \le j}\{\text{OPT}[i,r-1] + \text{OPT}[r+1,j]\}.$$

See table 6.9 for the root indices.

TABLE 6.9 The table root$[i, j]$ of root indices

	0	1	2	3	4	5
1		1	1	2	2	2
2			2	2	2	4
3				3	4	5
4					4	5
5						5

6.4 The Bellman-Ford Algorithm

In chapter 4.5 we explored Dijkstra's algorithm for finding the shortest paths from a single source vertex to all other vertices. Dijkstra's algorithm works only on graphs with nonnegative-weight edges. If some edge weights are negative, then Dijkstra's algorithm could return incorrect results. As an example, consider the graph in figure 6.6. Dijkstra's algorithm would visit vertex C first and return the distance 3. However, there is a shorter path S-A-B-C with the distance 1. Due to the greedy nature of the algorithm, the new distance to C won't be recorded.

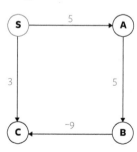

FIGURE 6.6 Dijkstra's algorithm does not work if there are negative edges.

You may think that there is an easy way to fix the algorithm by adding a large constant to each edge weight. Unfortunately, this idea does not work. That is because paths with more edges will be penalized disproportionately. If we add 9 to all edge weights in the previous graph and run Dijkstra's algorithm, the path S-A-B-C isn't the shortest anymore. This is illustrated in figure 6.7.

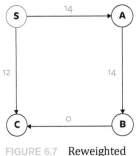

FIGURE 6.7 Reweighted graph.

There are two ideas to fix Dijkstra's algorithm: either to add a large constant to each path (Johnson's algorithm, which we won't cover in this book) or to relax all edges V-1 times (Bellman-Ford's algorithm). In this section we will consider the latter, since the former is not a dynamic programming algorithm. How can we use dynamic programming to find the shortest path? We need to somehow define *ordered* subproblems, otherwise we may get an exponential runtime. Consider the shortest v-u path $v = w_0, w_1, ..., w_{k-1}, w_k = u$. To have an optimal substructure the path $v = w_0, w_1, ..., w_{k-1}$ must be the shortest path from v to to w_{k-1}. Thus, we will be counting *the number of edges* in the shortest path. This is how we order subproblems.

Let $D[v, k]$ denote the length of the shortest path from s to v that uses at most k edges. How do we compute $D[v, k]$? By reducing it to subproblems of the smaller size. We can go to some neighbor w of v and then take the shortest path from s to w that uses at most $k - 1$ (which is already solved).

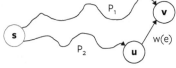

In figure 6.8 the paths $P_1 = D[v, k - 1]$ and $P_2 = D[u, k - 1]$ use at most $k - 1$ edges. The vertex v is adjacent to u. Then the path $P_3 = P_2 + (u, v)$ uses at most k edges and its length is $D[v, k] = w(u,v) + D[v, k - 1]$.

FIGURE 6.8 Defining subproblems.

Now there are two s-v paths: $P_1 = D[v, k - 1]$ and $P_3 = D[v, k]$. We do not know which path is actually shorter; therefore, we have to try both possibilities and then choose the minimum:

$$D[v,k] = \min_{(u,v)\in E} \{D[v,k-1], w(u,v)+D[v,k-1]\}.$$

This recursive definition is terminated by $D[v, 0] = 0$.

Here is a pseudocode:

```
D[v,0] = 0; for all v
for k = 1 to V-1:
  for each v in V:
    D[v, k] = D[v, k-1]
    for each edge (u,v)∈E
      D[v, k] = min(D[v, k-1], w(u, v) + D[u, k-1])
```

Note that the Bellman-Ford algorithm is designed only for directed graphs. For undirected graphs with negative-weight edges the shortest path problem is more complex and requires different algorithms.

Let us trace the algorithm on the example below (figure 6.9).

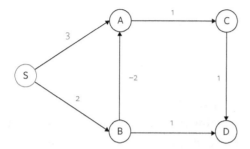

	A	B	C	D
D[v, 1]	3	2		
D[v, 2]	0	2	4	3
D[v, 3]	0	2	1	3
D[v, 4]	0	2	1	2

FIGURE 6.9   Tracing the algorithm.

Runtime. There are V^2 subproblems, and each subproblem takes $O(V)$ time to compute. Thus, the total running time is $O(V^3)$.

Note that the algorithm only finds the length of the shortest paths, but not the actual shortest paths. For that we need to store some axillary information. We create another array of vertices p[0 ... $V - 1$], where for each vertex v we store its predecessor in the shortest s-v path as in figure 6.10.

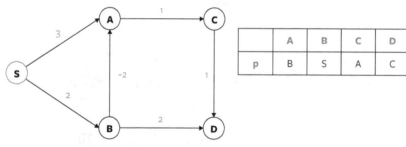

	A	B	C	D
p	B	S	A	C

FIGURE 6.10   Graph and its table of predecessors.

Having table of predecessors, we restore the path recursively. For example, to get the S-D path, we have to first get to p[D] = C, and then to p[C] = A, and then to p[A] = B, and finally to p[B] = S.

How would we apply the Bellman-Ford algorithm to find out if a graph has a negative cycle? Consider the following graph. The S-C distance is 3 if we take just an S-D edge. On the other hand, the distance is 1 if we take a path S-A-B-C. Moreover, the S-C distance is -1, if we take S-A-B-C-A-B-C path. The S-C distance can be as low as we want by going through a negative cycle C-A-B-C. This tells us that the shortest path problem does not have a solution in presence of a negative cycle.

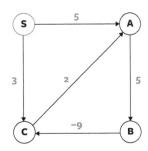

FIGURE 6.11   A graph with a negative weight cycle.

However, the Bellman-Ford algorithm can easily detect if a graph has a negative cycle. The procedure is the following: Do not stop after $V - 1$ iterations, perform one extra round, and if anything changes in the table, then we know there is a negative cycle.

6.5   The Shortest Path in DAGs

In this chapter we will solve a shortest distance problem in weighed directed acyclic graphs (DAG). For these special graphs we will develop a dynamic programming algorithm that is faster than the Bellman-Ford algorithm from the previous chapter and the Dijkstra algorithm from chapter 4.5. We do not require edge weights to be nonnegative and we don't have to worry about negative-weight cycles, since a DAG is acyclic.

Recall a topological sort from chapter 1.3.2. If graph $G = (V, E)$ is a DAG, it is always possible to arrange vertices in a topological order. The runtime complexity of the algorithm is linear $O(V + E)$. Figure 6.12 demonstrates a DAG and one possible ordering.

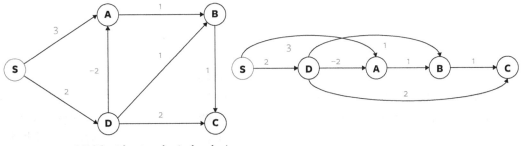

FIGURE 6.12   A DAG with a topological ordering.

You can see from the picture that whenever we have an edge from u to v, the ordering visits u before v. Therefore, in the dynamic programming approach we organize subproblems according to the topological ordering. We will pass through the ordered list and compute distances just like in Dijkstra's algorithm. Let $d(v)$ denote the length of the shortest path from s to v for each $v \in V$. We compute $d(v)$ as the minimum over all adjacent vertices:

$$d(v) = \min_{(u,v) \in E} \{d(u) + w(u,v)\}.$$

Note that vertex u is preceding vertex v. In figure 6.12, $d(C) = \min(d(B) + 1, d(D) + 2)$. Here is a pseudocode:

```
d[s] = 0, d[v] = infinity for all v ∈ V\{s}
topologically sort the vertices
for each v taken in topological order
    for each u ∈ adjacent[v]
        if d[v] > d[u] + w(u, v) then d[v] = d[u] + w(u, v)
```

The runtime complexity is $\Theta(V + E)$, since it requires a single pass over vertices in topological ordering and relaxing each edge that leaves each vertex. As an example, we run the algorithm over a graph from figure 6.12.

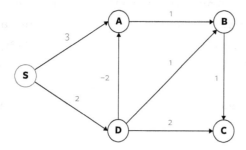

	A	B	C	D
S	3	∞	∞	2
D	0	3	4	
A		1		
B			2	

FIGURE 6.13   Steps of the algorithm.

In figure 6.13 we show a table $d[v]$, where each row represents a vertex in a topological ordering S-D-A-B-C. The table is filled in row-by-row fashion. In each table entry (i, j) we record an updated distance from the vertex s to vertex j via an adjacent vertex i. If the table entry is empty, then the distance was not updated. The algorithm does not find the actual shortest paths, but only calculates the distances. As with the Bellman-Ford algorithm, we can add an array $p[]$ such that $p[v]$ stores the vertex previous to v in the shortest path from s to v. This will allow us to reconstruct the actual shortest paths.

1. (**T/F**) If a dynamic programming algorithm has n subproblems, then its running time complexity is $\Omega(n)$.

2. (**T/F**) It is possible for a dynamic programming algorithm to have an exponential runtime complexity.

3. (**T/F**) In a dynamic programming formulation, the subproblems must be mutually independent.

4. (**T/F**) A pseudo-polynomial time algorithm is always asymptotically slower than a polynomial time algorithm.

5. (**T/F**) If a dynamic programming solution is set up correctly (i.e., the recurrence equation is correct) and each unique sub-problem is solved only once, then the resulting algorithm will always find the optimal solution in polynomial time.

6. (**T/F**) If a problem can be solved by divide and conquer, then it can always be solved by dynamic programming.

7. (**T/F**) If a problem can be solved by dynamic programming, then it can always be solved by exhaustive search.

8. (**T/F**) The Bellman-Ford algorithm always fails to find the shortest path between two nodes in a graph if there is a negative cycle present in the graph.

9. (**T/F**) In a dynamic programming solution, the space requirement is always at least as big as the number of unique sub problems.

10. (**T/F**) In a connected, directed graph with positive edge weights, the Bellman-Ford algorithm runs asymptotically faster than the Dijkstra algorithm.

11. (**T/F**) The dynamic programming for the knapsack problem runs in polynomial time.

12. (**T/F**) The longest simple path can be computed by negating the weights of all the edges in the graph and then running the Bellman-Ford algorithm.

13. (**T/F**) There exist some problems that can be solved by dynamic programming but cannot be solved by greedy algorithms.

14. (**T/F**) The Bellman-Ford algorithm always finds the shortest path in undirected graphs.

15. Which of the following standard algorithms are solved using dynamic programming?

    a. Bellman-Ford's algorithm

    b. Dijkstra's algorithm

    c. Prim's algorithm

    d. Karatsuba's algorithm

EXERCISES

1. Design a DP algorithm that solves the 0-1 knapsack problem, which allows repetitions (i.e., assume that there are unlimited quantities of each item available). What is its space complexity?

2. Design a DP algorithm that takes a string and returns the length of the longest palindromic subsequence. A subsequence of a string is obtained by deleting zero or more symbols from that string. A subsequence is palindromic if it reads the same left and right. For example, the string QRAECCETCAURP has several palindromic subsequences, but the longest one is RACECAR.

3. Given a non-empty string *str* and a dictionary containing a list of unique words, design a dynamic programming algorithm to determine if *str* can be segmented into a sequence of dictionary words. For example, if *str* = "algorithmdesign" and your dictionary contains "algorithm" and "design," then your algorithm should answer yes since *str* can be segmented to "algorithm" and "design." You may assume that a dictionary lookup can be done in $O(1)$ time.

4. You are given n balloons, indexed from 0 to $n - 1$, where each balloon is painted with a number on it represented by array *nums*. You are asked to burst all the balloons. If you burst balloon i you will get *nums*[left] · *nums*[i] · *nums*[right] coins. Here, left and right are adjacent indices of i. After the burst, the left and right then becomes adjacent. You may assume *nums*[−1] = *nums*[n] = 1, and they are not real; therefore, you cannot burst them. For example, if you have the *nums* = [3, 1, 5, 8], the optimal solution would be 167, where you burst balloons in the order of 1, 5, 3 and 8. The array *nums* after each step is [3, 1, 5, 8] → [3, 5, 8] → [3, 8] → [8] → []. Design a dynamic programming algorithm to find the maximum coins you can collect by bursting the balloons. Analyze the running time of your algorithm.

5. A rope has length of n units, where n is an integer. You are asked to cut the rope (at least once) into different smaller pieces p_j of integer lengths so that the product of lengths of those new smaller ropes is maximized. Design a dynamic programming algorithm and analyze its running time. Explain how you would find the optimal set of cutting positions.

6. There is a series of $n > 0$ jobs lined up one after the other. The *i-th* job has a duration $t_i \in \mathbb{N}$ units of time, and you earn $p_i \geq 0$ amount of money for doing it. Also, you are given the number $s_i \in \mathbb{N}$ of immediately following jobs that you cannot take if you perform that *i-th* job. Design a dynamic programming algorithm to maximize the amount of money one can make in T units of time.

7. You are to compute the minimum number of coins needed to make change for a given amount m. Assume that we have an unlimited supply of coins. All denominations d_k are sorted in ascending order: $1 = d_1 < d_2 < ... < d_n$. Design a dynamic programming algorithm to minimize the amount of coins.

8. Given an unlimited supply of coins of denominations $d_1 < d_2 < ... < d_n$, we wish to make change for an amount m. This might not be always possible. Your goal is to verify if it is possible to make such change. Design an algorithm by reduction to the knapsack problem.

9. There are two strings: string S of length n, and string T of length m. Design a dynamic programming algorithm to compute their longest common subsequence. A subsequence is a subset of elements in the sequence taken in order (with strictly increasing indexes.)

10. A polygon is called convex if all its internal angles are less than 180°. A convex polygon is represented as an array V with n vertices in counterclockwise order, where each vertex is in the form of a coordinate pair (x, y). Given is a convex polygon, we would like to triangulate this polygon (i.e., decompose it into disjoint triangles by adding line segments (diagonals) between its corners (vertices)). Design a dynamic programming algorithm for triangulating a convex polygon while minimizing the total perimeter of all the triangles.

11. Given a row of n houses that can each be painted red, green, or blue with a cost $P(i, c)$ for painting house i with color c, design a dynamic programming algorithm to find a minimum cost coloring of the entire row of houses such that no two adjacent houses are the same color.

12. A tourism company is providing boat tours on a river with n consecutive segments. According to previous experience, the profit they can make by providing boat tours on segment i is known as a_i. Here, a_i could be positive (they earn money), negative (they lose money), or zero. Because of the administration convenience, the local community requires that the tourism company do their boat tour business on a contiguous sequence of the river segments (i.e., if the company chooses segment i as the starting segment and segment j as the ending segment, all the segments in between should also be covered by the tour service, no matter whether the company will earn or lose money). The company's goal is to determine the starting segment and ending segment of boat tours along the river, such that their total profit can be maximized. Design a dynamic programming algorithm to achieve this goal and analyze its runtime.

13. You have two rooms to rent out. There are n customers interested in renting the rooms. The ith customer wishes to rent one room (either room you have) for $d[i]$ days and is willing to pay $bid[i]$ for the entire stay. Customer requests are nonnegotiable in that they would not be willing to rent for a shorter or longer duration. Design a dynamic programming algorithm to determine the maximum profit that you can make from the customers over a period of D days.

14. You are to plan the fall 2025 schedule of classes. Suppose that you can sign up for as many classes as you want, and you'll have infinite amount of energy to handle all the classes, but you cannot take two classes at the same time. Also assume that the problem reduces to planning your schedule for one particular day. Thus, consider one day of the week and all the classes happening on that day: $c_1, ..., c_n$. Associated with each class c_i is a start time s_i and a finish time f_i such that $s_i < f_i$. Also, there is a score v_i assigned to that class, c_i, based on your interests and your program requirement. You would like to choose a set of courses for that day to maximize the total score. Design a dynamic programming algorithm for planning your schedule.

15. There are n trading posts along a river numbered $n, n-1 ..., 1$. At any of the posts you can rent a canoe to be returned at any other post downstream. (It is impossible to paddle against the river.) For each possible departure point i and each possible arrival point $j < i$, the cost of a rental is $C[i, j]$. However, it can happen that the cost of renting from i to j is higher than the total costs of a series of shorter rentals. In this case you can return the first canoe at some +post k between i and j and continue your journey in a second (and, maybe, third, fourth, and so on) canoe. There is no extra charge for changing canoes in this way. Design a dynamic programming algorithm to determine the minimum cost of a trip by canoe from each possible departure point i to each possible arrival point j. Analyze the running time of your algorithm in terms of n.

16. Given a weighted directed acyclic graph $G = (V, E)$ in which we allow negative edge weights, design a dynamic programming algorithm to find the longest simple path between two given vertices.

17. Design a dynamic programming algorithm for counting the number of paths between two given vertices in a DAG.

Chapter 7

Network Flow

I N THIS CHAPTER WE WILL LEARN our fourth major algorithm design technique (after greedy, divide-and-conquer, and dynamic programming). Network flow is an important design technique because it can be used to express a wide variety of problems. When we think of networks, we typically envision a physical network, like an electrical network (with an electrical current flow), or a hydraulic network (with a water, gas, or oil flow), or a communication network (with a voice, data, or video flow), or a transportation network (with passengers, vehicles, or freight flow). Transportation networks are the most popular; they are designed to model complex distribution and logistics decisions. In this model, a shipper with an inventory of goods at its warehouses must ship to disperse retail centers (with different customer demands) given transportation routes. Each route has a distribution capacity and cost. The goal is to ship the maximum amount of goods.

7.1   Introduction

We start with a directed weighted graph $G = (V, E)$ with two distinguished vertices s (the source) and t (the sink), in which each edge $(u, v) \in E$ has a nonnegative capacity $c(u, v)$. The graph should never have edges between u and v in both directions, so there are no loops. Also, if $u, v \in V$ but $(u, v) \notin E$, we assume that $c(u, v) = 0$. We call this graph a *flow network*. Next, we define a *flow* as a function f that assigns nonnegative real values to the edges of G and satisfies two axioms:

1. Capacity constraint: $0 \leq f(u, v) \leq c(u, v)$, for each $u, v \in V$
2. Conservation constraint: $\sum_u f(u,v) = \sum_w f(v,w)$, for each $v \in V - \{s, t\}$

In other words, the flow does not exceed the capacity on any edge, and the flow entering a vertex equals the flow leaving the vertex at every vertex other than the source and the sink.

We also define a *value* of the flow: the total flow that the source s can send, $|f| = \Sigma_v f(s,v)$. Since s and t are the only nodes that are not beholden to the conservation law, the value of f can be equivalently stated as the amount of flow entering t.

The max-flow problem is stated as to find the maximum flow value into the target t.

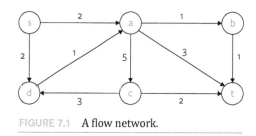

FIGURE 7.1   A flow network.

In the graph in figure 7.1, $|f| = 4$, however, the max flow is only 3; we push 2 units of flow along the edge (s, a), and one unit of flow along the edge (s, d). How do we prove that this is the max flow? The flow saturates edges (s, a) and (d, a). If we remove them, the graph becomes disconnected.

Let us consider a greedy approach to the max-flow problem: choosing an edge leaving the source with the largest capacity. This greedy algorithm does not find the max flow in general graphs. A simple counterexample can be seen in figure 7.2.

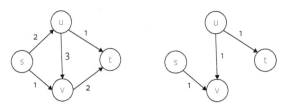

FIGURE 7.2   Pushing 2 units of flow via *s-u-v-t*.

In figure 7.2, the greedy algorithm has made a first choice to push 2 units of flow through *s-u-v-t* path—the maximum flow has not been achieved. The optimal flow value is 3: We push one unit via *s-u-t* and then another unit via *s-v-t*, and one more unit via *s-u-v-t*. The problem with a greedy approach is that we pushed too much flow via the (u, v) edge. We want to redo our previously made decision and push on that edge only one unit of flow. Unfortunately, a greedy approach does not allow us to change the previously made decisions. But what if we can push a unit of flow back through (u, v)? This will mean that we cancel the previously pushed flow by one unit. This is the rough idea of the Ford–Fulkerson algorithm. We modify the greedy algorithm such that we can revise the paths later by flow cancelation. Thus, there are two ways to increase a flow value:

- Find unused capacity

- Find cancelable flow

We will keep track of how much additional flow can be pushed directly (over an edge) between any pair of vertices u and v (in each direction). This requires constructing another directed graph G_f, called the *residual network* of f, which has the same vertices as G, but a different set of edges E_f. Assume a flow network with some flow f on each edge. Then, for each edge $(u, v) \in E$ we create

- a forward edge, and we include edge (u, v) into G_f with the residual capacity $c_f(u, v) = c(u, v) - f(u, v)$; and

- a backward edge, and we include edge (v, u) into G_f with the residual capacity $c_f(v, u) = f(u, v)$.

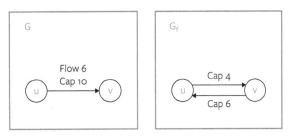

FIGURE 7.3   Example of residual capacities.

Having backward edges allows us to fix the greedy approach by erasing a flow on some edges. Next, we define an *augmenting path*. Let P be a simple (with no cycles) path from s to t in G_f. We can find such a path by running a graph traversal. The residual capacity of P is the smallest capacity on any edge of P, namely $c_f(P) = \min\{c_f(u, v) : (u, v) \in P\}$. If $c_f(P) > 0$, then P is an *augmenting path* in G_f.

As an example, consider the graph G in figure 7.4. Suppose we push two units of flow on s-d-b-t path. We will end up with the residual graph G_f. Note that edge (d, b) is saturated, $c_f(d, b) = 0$; we do not include that edge into G_f.

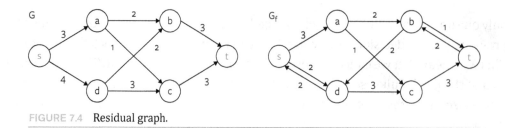

FIGURE 7.4 Residual graph.

In G_f we can find another augmenting path, for example, s-a-b-d-c-t, and push two units of flow along the path.

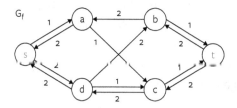

FIGURE 7.5 Augmenting flow along s-a-b-d-c-t path.

Note in this path we erased the previous flow on (d, b) edge. From this example we see that in the residual network $G_f = (V, E_f)$ we can increase the flow by using forward as well as backward edges as long as there is an augmenting path. The residual network and augmenting along an s-t path are the cornerstone of Ford-Fulkerson algorithm.

7.2   The Ford–Fulkerson Algorithm

For the purpose of this algorithm, we will assume that all capacities and all flows take only nonnegative, integral values. The algorithm begins with the zero flow f and successively improves f by finding an augmenting s-t path P and pushing as much flow as possible along the path. It terminates if there are no more s-t paths in G_f. The Ford–Fulkerson algorithm is essentially a greedy algorithm; it finds a locally optimal solution which turns out to be a global optimum. Here is a pseudocode for the algorithm:

```
Given a flow network: (G=(V, E), s, t, c)
1.   start with f(u, v)=0 and G_f = G //initialization
2.   while (there exist an augmenting path P in G_f):
3.       find a bottleneck c_f(P) = min{c_f (u, v): (u, v) ∈ P}.
4.       augment the flow f along P
5.       update the residual graph G_f
```

7.2.1 Example

Let us run Ford–Fulkerson's algorithm on the graph G in figure 7.6.

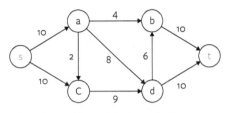

FIGURE 7.6   The original graph G.

We will illustrate iterations of the algorithm on the residual graph G_f. There will be multiple augmenting paths in G_f, so we will make an arbitrary choice. There are many heuristics for choosing an augmenting path, which we will address later. We start with a zero flow and $G_f = G$. We find an augmenting path s-a-d-t with the bottleneck 8. We push 8 units of flow and augment the flow along that path and update the residual graph as in figure 7.7.

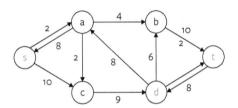

FIGURE 7.7   Residual graph G_f after pushing 8 units.

Next, we find another augmenting path s-a-c-d-t in G_f with the bottleneck 2. We push 2 units of flow and augment the flow along that path and update the residual graph as in figure 7.8. The total flow now is 10.

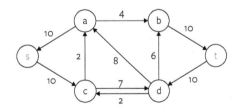

FIGURE 7.8   Residual graph G_f after second iteration.

Then again, we find an augmenting path *s-c-a-b-t* with the bottleneck 2. We push 2 units of flow and augment the flow along that path and update the residual graph as in figure 7.9. The total flow now is 12.

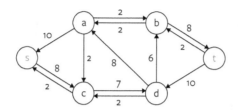

FIGURE 7.9   Residual graph G_f after third iteration.

On the next iteration we pick an augmenting path *s-c-d-b-t* with the bottleneck 6. The updated residual graph is depicted in figure 7.10. The total flow now is 18.

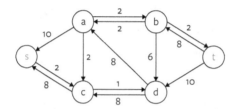

FIGURE 7.10   Residual graph G_f after fourth iteration.

On the fifth iteration we pick an augmenting path *s-c-d-a-b-t* with the bottleneck 1. The updated residual graph is depicted in figure 7.11. The total flow now is 19.

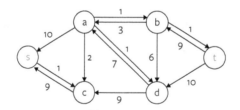

FIGURE 7.11   Residual graph G_f after fifth iteration.

That was the algorithm's last iteration. As you easily see from figure 7.11 the residual graph is disconnected—there is no an *s-t* path in it. Therefore, we found the maximum flow of 19 units. In figure 7.12 we demonstrate the original network flow graph *G* with each edge labeled by flow/capacity.

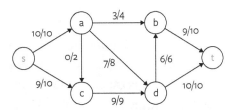

FIGURE 7.12   Network flow graph with flow/capacity on each edge.

7.2.2   Complexity of the Ford–Fulkerson Algorithm

We find an augmenting path (line 2 in the pseudocode) through a graph traversal in $O(E)$ time (since the number of edges in G_f is at most $2E$). In each step of the algorithm we traverse the path to find a bottleneck (line 3) and traverse it again to update the residual graph (line 5). These also take $O(E)$ time. The question remains, "How many steps are in the while loop (lines 2–5)?" Since the edge capacities are integral, the bottleneck = $\min \{c_f(u, v): (u, v) \in P\}$ is also integral. It follows that in the worst case we increase the value of flow by at least one. Hence, the algorithm stops after at most $|f| = \Sigma_v f(s, v)$ steps. This implies that the running time of the Ford–Fulkerson algorithm is $O(E \cdot |f|)$ for integral capacities. The algorithm is pseudo-polynomial (see Chapter 6.2.1) because it depends on the size of the integers in the input.

The following example demonstrates an extreme case of the algorithm's slow convergence. Consider a graph where four edges have capacities of $c = 10^9$ and one edge has capacity of 1. On each iteration we choose an augmenting path in such a way that nodes u and v are always in the path.

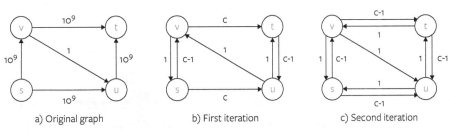

a) Original graph   b) First iteration   c) Second iteration

FIGURE 7.13   Extreme case of the Ford–Fulkerson algorithm.

Since each iteration increases the flow value by 1, the algorithm terminates after $2 \cdot 10^9$ steps.

Note that the algorithm may never terminate if the edge capacities are arbitrary real numbers. The algorithm can loop forever, always finding smaller and smaller augmenting paths. See the example of this effect by U. Zwick.[1]

---

1   "Ford–Fulkerson Algorithm," Wikipedia, https://en.wikipedia.org/wiki/Ford–Fulkerson_algorithm#Non-terminating_example

7.2.3 Proof of Correctness

When no more augmenting paths can be found, the graph becomes disconnected and therefore no more flow can be pushed from s to t in the residual network. This proves that the flow we found is maximal. Is this the maximum? Maximum is not the same as maximal. Since we choose the augmenting paths arbitrarily, it seems it may happen that when running the algorithm for the second time we will get a bigger flow. We prove the maximum flow by using the duality principle. An optimization problem may be viewed from two perspectives, the *primal* (minimization) problem or the *dual* (maximization) problem. The solution to the dual problem provides a lower bound to the solution of the primal problem. We think of the maximum flow problem as the dual problem. We will formulate the primal problem in terms of a vertex cut.

A vertex s-t cut of a flow network is a partition of the vertices V into disjoint subsets A and B such that $s \in A$ and $t \in B$. We define the cut capacity, cap(A, B), as the sum of capacities of all the edges going from partition A to partition B. In figure 7.14, partition A consists of vertices s, a, b, and partition B consists of vertices t, c, d. The cut capacity is cap(A, B) $= 10 + 2 + 8 + 10 = 30$.

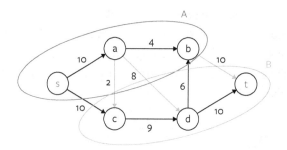

FIGURE 7.14   *st*-cut and its capacity.

The minimum cut problem is to compute an s-t cut whose capacity is as small as possible. We will show that the value of any flow is at most the capacity of any cut.

Lemma 1. *For any flow and any cut:*

$$|f| = \sum_{v} f(s,v) = \sum_{u \in A, v \in B} f(u,v) - \sum_{u \in A, v \in B} f(v,u).$$

Proof. Since there are no incoming edges to s, we have that $f(v, s) = 0$, and therefore

$$|f| = \sum_v f(s,v) = \sum_v f(s,v) - \sum_v f(v,s).$$

Next, we observe that due to the flow conservation law

$$\sum_v f(u,v) = \sum_v f(v,u)$$

for any vertex u except s and t. It follows,

$$|f| = \sum_{u \in A} \left(\sum_v f(u,v) - \sum_u f(v,u) \right) = \sum_{u \in A,\, v \in B} f(u,v) - \sum_{u \in A,\, v \in B} f(v,u).$$

This concludes the proof. ∎

The graph in figure 7.15 illustrates Lemma 1.

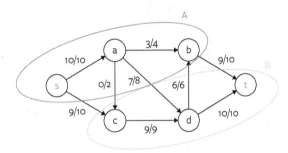

FIGURE 7.15   This demonstrates Lemma 1.

The flow from the source $|f|$ is 19. The flow from partition A to B is $9 + 7 + 9 = 25$. The flow from partition B to A is 6. The flow difference is 19, same as $|f|$.

Lemma 2. *For any flow and any (A, B)- cut:*

$$|f| \leq cap(A, B).$$

Proof. By previous Lemma 1 and taking into account the capacity constraint, we obtain

$$|f| = \sum_{u \in A,\, v \in B} f(u,v) - \sum_{u \in A,\, v \in B} f(v,u) \leq \sum_{u \in A,\, v \in B} f(u,v).$$

Taking into account the capacity constraint, we obtain

$$|f| \le \sum_{u \in A,\, v \in B} f(u,v) \le \sum_{u \in A,\, v \in B} c(u,v) = cap(A,B).$$

This concludes the proof. ∎

Lemma 2 proved that the solution to the min-cut problem provides an upper bound to the solution of the max-flow problem:

$$\max_{f} |f| \le \min_{(A,B)} cap(A,B).$$

In fact, this bound is tight.

Theorem 1. *The Ford–Fulkerson algorithm outputs the maximum flow.*

Proof. When the algorithm terminates there is no augmenting path from s to t in the residual graph G_f. Let A be a set of vertices reachable from s in G_f and let set B be all other vertices in V including t. We will prove that $|f| = cap(A, B)$. Consider any edge (u, v) from A to B in the original flow network. This edge cannot exist in G_f, because otherwise vertex v will be reachable from s, which contradicts the definition of the s-t cut. It follows that that edge must be saturated $f(u, v) = c(u, v)$ in G_f. Now consider any edge (v, u) from B to A in the original flow network. The flow on this edge must be zero $f(v, u) = 0$. If $f(v, u) > 0$, then there will be an edge in the opposite direction (u, v) in G_f, and therefore vertex v will be reachable from s. Again, we reached a contradiction to the definition of s-t cut. Then by Lemma 1, we have

$$|f| = \sum_{u \in A,\, v \in B} f(u,v) - \sum_{u \in A,\, v \in B} f(v,u) = \sum_{u \in A,\, v \in B} c(u,v) - 0 = cap(A,B).$$

Thus, the Ford–Fulkerson algorithm outputs the maximum flow, the cut (A, B) is a minimum cut, and the max-flow equals the capacity of the min-cut. ∎

7.3 Reduction to Network Flow

A reduction is a problem-solving method for transforming instances of problem Y into instances of another problem X, so that an algorithm for solving problem X efficiently can be used to solve problem Y efficiently. Formally, to reduce a problem Y to a problem X (we write $Y_p \leq X$) we want a function f that maps Y to X such that

1. f is a polynomial time computable and
2. $\forall$ instance $y \in Y$ is solvable if and only if $f(y) \in X$ is solvable.

Figure 7.16 illustrates the idea of problem solving by polynomial-time reduction.

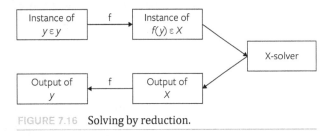

FIGURE 7.16   Solving by reduction.

If problem X can be solved in polynomial time and $Y_p \leq X$, then Y can be solved in polynomial time. This is the most common use of reductions. In chapter 9 we will see that reductions also can be used to prove that problem X is $\mathcal{NP}$-hard.

7.3.1   Dinner Party

This is our first example of solving a problem by using a reduction to network flow.

> At a dinner party, there are n families $f_1, f_2, ..., f_n$ and m tables $t_1, t_2, ..., t_m$. The i-th family f_i has r_i relatives and the j-th table b_j has s_j seats. Everyone is interested in making new friends between families; therefore, the dinner party planner wants to seat people such that no two members of the same family are seated at the same table. Design an algorithm that decides if there exists a seating assignment such that everyone is seated and no two members of the same family are seated at the same table. What would be a seating arrangement?

We start by setting the problem as a bipartite graph problem. In this graph one partition is a set of vertices, f_i, representing all n families. Another partition is a set of m tables t_j. Then, we connect each family f_i to all tables t_j by directed edges with the capacity 1.

Next, we extend the bipartite graph to a network flow. We add the source s and connect it to every family vertex f_i by an edge (s, f_i) of capacity r_i. We add the target t and for every table vertex t_j, we add an edge (t_j, t) of capacity s_j.

Claim. *The original problem has a solution (a valid seating assignment is indeed possible) if and only if the constructed network has a max-flow of value $r_1 + r_2 + ... + r_n$.*

Proof. $\Rightarrow$) Assume that there is a solution. It means that every family member is seated. So, we can push a flow of r_i from the source s to each family. On the edges between families and tables, we assign a flow of 1 or 0. Since no two members of the same family are seated at the same table, each family vertex will have outgoing flow of value 1. On the edges between tables and the sink, we assign a flow value equal to the number of people seated at that table. This must be possible, since we have a valid assignment.

Conversely $\Leftarrow$) Assume there is a max-flow of value $r_1 + r_2 + ... + r_n$. This means that each family vertex will get a flow of r_i. Due to capacity constrain (each edge (f_i, t_j) has a unit capacity) no two members will sit at the same table. We also observe that no table is overloaded due to the capacity condition s_i. ∎

Lastly, we get a seating assignment by running a network flow algorithm and pick edges (f_i, t_j) with a unit flow.

7.3.2 Reallocation Problem

As the second example of using a reduction to network flow, we consider the following problem:

> A company has n locations in city A and plans to move some of them (or all) to another city B. The i-th location costs a_i per year if it is in the city A and b_i per year if it is in the city B. The company also needs to pay an extra cost, $c_{ij} > 0$, per year for traveling between locations i and j. We assume that $c_{ij} = c_{ji}$. Design an efficient algorithm to decide which company locations in city A should be moved to city B in order to minimize the total annual cost.

We start with constructing a flow network. Create a complete graph where each vertex $v_i, i = 1, 2, ..., n$ is a company location in city A. Any two vertices v_i and v_j are connected by a bidirectional edge with capacity $c_{ij} > 0$. We connect the source s to all vertices $v_i, i = 1, 2, ..., n$ with capacity b_i on edge (s, v_i). Finally, we connect all vertices $v_i, i = 1, 2, ..., n$ to the sink t with capacity a_i on edge (v_i, t).

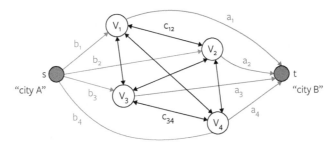

FIGURE 7.17   Flow network for 4 cities.

We have constructed a flow network with $V = n + 2$ vertices and $E = 2n + n$ $(n-1)/2$ edges. Figure 7.17 demonstrates a flow network of four cities. Next, we run the Ford–Fulkerson algorithm (see Exercise 1) to separate all vertices into two partitions in such a way that the cut capacity is the smallest. See figure 7.18 for a possible min-cut.

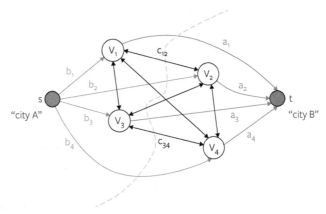

FIGURE 7.18   A min-cut for 4 cities.

The min-cut in figure 7.18 suggests that locations v_2 and v_4 should be moved to city B. Noting that max flow is $a_1 + a_3 + b_2 + b_4 + c_{12} + c_{14} + c_{32} + c_{34}$, we propose the following claim.

Claim. *The total annual cost is minimized if and only if the constructed flow network has a max flow of the following value*

$$\sum_{i \in A} a_i + \sum_{j \in B} b_j + \sum_{i \in A, j \in B} c_{ij}.$$

Proof. ⇒) The max flow is exactly the annual cost of moving the locations from city A to city B. Thus, the minimum cut of this network corresponds to the optimal solution of the relocation problem.

Conversely ⇐) Consider the network with the max flow. By the property of the min-cut,

- the blue edges with capacity b_j are saturated, meaning that the location v_j is moved to city B;

- the red edges with capacity a_i are saturated, meaning that the location v_i stays in city A; and

- the black edges with capacity c_{ij} are saturated, meaning that the cost of *for* traveling between locations v_i and v_j.

The runtime complexity, assuming the Ford–Fulkerson algorithm, is $O(E \cdot |f|) = O(n^2 \cdot |f|)$. ∎

7.4 Augmenting Path Heuristics

We have seen in figure 7.13 that the way augmenting paths are chosen can significantly impact the algorithm's performance. Here, we consider a couple of heuristics (due to Jack Edmonds and Richard Karp) of selecting augmenting paths to avoid that extreme performance of the Ford–Fulkerson algorithm.

One suggestion is that we should select edges with high capacities, called the "maximum bottleneck path." In the residual network in figure 7.13, the paths *s-v-t* and *s-u-t* have weight 10^9, while the path *s-v-u-t* has weight 1.

Another suggestion is that we should find the shortest augmenting path in terms of the number of edges. This approach does not consider the edge capacities at all. In the residual network in figure 7.13, the paths *s-v-t* and *s-u-t* are one edge shorter than the path *s-v-u-t*.

There are many other augmenting path heuristics for the Ford–Fulkerson algorithm.

7.4.1 Edmonds–Karp 1: Augmenting Path with Largest Capacity

We consider an implementation of the Ford–Fulkerson algorithm in which we pick the augmenting path with the largest bottleneck value. In this scenario, we need to repeatedly find the path between two vertices whose minimum capacity is the largest. That path can be found using an algorithm similar to Dijkstra's shortest-path algorithm. Instead of maintaining the shortest path length to a vertex, we maintain the bottleneck. Let bottleneck(v) be the capacity of the highest-capacity path from s to v. In this array we will keep a track of the lowest capacity edge on s-v path discovered so far. We will also maintain a spanning tree T of vertices, rooted at s, for which we have bottleneck(v). If we find another s-v path with a higher bottleneck value, we update bottleneck(v). Here is the updated rule:

$$\text{bottleneck}(v) = \max_{\substack{u \in T \\ (u,v) \in E}} \{\min(\text{bottleneck}(u), c(u,v))\}$$

and a pseudocode for finding the largest-capacity path:

```
while T ≠ V
    for each v ∈ V adjacent to T:
        update bottleneck(v);
    add v to T;
end
```

That path can be found in $O(E \cdot \log V)$ time using a binary max heap. The runtime analysis is the same as in Dijkstra's algorithm.

So far, we have addressed the runtime complexity of a single iteration in the Edmonds–Karp algorithm. Next, we compute the upper bound on the total number of iterations in terms of the value of the maximum flow. In the Ford–Fulkerson algorithm we increase the flow by any path bottleneck (which could be as low as just 1); in the Edmonds–Karp algorithm we increase the flow by the maximum path bottleneck.

Claim 1. *If the max flow in the network is $|f|$, then there exists an s-t path with capacity of $\geq |f|/E$.*

Proof. To prove the existence of such a path, we delete all edges with capacity $< |f|/E$. Let us call this graph G'. We claim that G' is <u>not</u> disconnected and has an s-t augmenting path. Suppose that G' is disconnected. Then, every edge on an s-t cut has capacity $< |f|/E$. Since in the worst case there could be E edges on that cut, it follows that the cut capacity is $cap(A, B) < E \cdot |f|/E = |f|$. This is a contradiction, since Lemma 2 says $|f| \leq cap(A, B)$. ∎

Claim 2. *Edmonds–Karp makes $O(E \cdot \log |f|)$ iterations.*

Proof. The previous claim says that each iteration adds at least $1/E$ fraction of the flow found so far. But let us run the algorithm backward. On each iteration the flow F gets reduced by at least $|f|/E$. So, after the first step the max in the residual network is at most

$$|f|-|f|\cdot\frac{1}{E}=|f|\cdot\left(1-\frac{1}{E}\right).$$

After the second step, the flow is at most

$$|f|\cdot\left(1-\frac{1}{E}\right)-|f|\cdot\left(1-\frac{1}{E}\right)\cdot\frac{1}{E}=|f|\cdot\left(1-\frac{1}{E}\right)^2.$$

Proceeding in the same way, after x steps, the max flow is at least $|f|\cdot(1-1/E)^x$. How many iterations x do we need to have in order to reduce the max-flow $|f|$ to 1? To answer that we need to solve the following inequality:

$$|f|\cdot\left(1-\frac{1}{E}\right)^x\leq1.$$

Noticing that $1+z\leq e^z$ (where e is the Euler constant), we find that $x=O(E\cdot\log|f|)$. This says that the flow decreases exponentially with the number of iterations. ■

We conclude that for graphs with integer capacities, the Edmonds-Karp 1 algorithm runs in $O(E^2\cdot\log V\cdot\log|f|)$ time.

As an example, let us consider a graph in figure 7.6 and run a few iterations of the above algorithm. The first augmenting path we choose is s-c-d-t. We push 9 units of flow and augment the flow along the path.

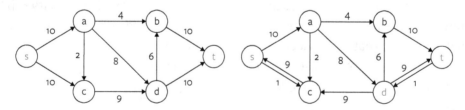

FIGURE 7.19   The residual graph after first iteration.

The next augmenting path with the largest bottleneck is *s-a-d-b-t*. We push 6 units of flow and augment the flow along the path.

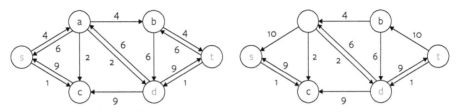

FIGURE 7.20   The residual graph after second iteration.

The last augmenting path is *s-a-b-t*. Comparing this algorithm with the Ford–Fulkerson algorithm, we see that we reached the max flow in only three iterations.

7.4.2   Edmonds–Karp 2: Shortest Augmenting Path

In this heuristic, we repeatedly select the shortest augmenting path, in terms of the number of edges. The resulting network flow algorithm is known as the Edmonds–Karp 2 algorithm. The shortest path can be found in $O(E)$ time by running a breadth-first search in the residual graph. The subtle question is, "How many iterations does the algorithm take?" It can be shown that this requires only $O(V \cdot E)$ iterations. Thus, the total runtime is $O(V \cdot E^2)$. The proof is quite elaborate and beyond the scope of this book.

7.5   The Circulation Problem

In this section we modify and extend the network flow problem, but this time there will be no source and sink. Also, we add demand $d(v)$ on each vertex and the lower bounds on the capacities on the given edges. This leads to the notion of *circulations* on graphs.

7.5.1   Circulation with Demands

Given a directed graph, in addition to having capacities $c(u, v) \geq 0$ on each edge, we associate each vertex v with a supply/demand value $d(v)$. We say that a vertex v is a demand if $d(v) > 0$ and a supply if $d(v) < 0$. If $d(v) = 0$ then the vertex simply receives and transmits flow.

The demand function $d(v)$ describes how much of an excess flow must be injected or extracted at each vertex. Next, we define a *circulation* with demands as a function f that assigns nonnegative real values to the edges of G and satisfies the following two axioms:

1.   Capacity constraint: $0 \leq f(u, v) \leq c(u, v)$, for each $u, v \in V$
2.   Conservation constraint: $\sum_u f(u,v) - \sum_w f(v,w) = d(v)$, for each $v \in V$

See figure 7.21 for an example, in which supply vertex *a* must send 3 units of flow and demand vertex *b* must receive 4 units of flow.

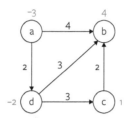

FIGURE 7.21   Circulation with demands (in red).

We call a circulation *feasible* if it meets the capacity and demand constraints. The circulation problem is stated as to find a feasible circulation. First, we note, that if there is a feasible circulation, then $\sum_v d(v) = 0$. We prove this by taking the conservation constraints and summing them up over all vertices:

$$\sum_v \left(\sum_u f(u,v) - \sum_w f(v,w) \right) = \sum_v d(v).$$

The left-hand side of the equality is zero, since the flow on every edge is summed twice, once as a coming-in flow, and then as a coming-out flow. This implies the claim. See figure 7.22.

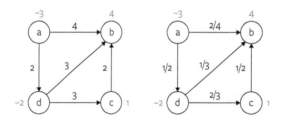

FIGURE 7.22   Left: Circulation with demands; right: Feasible circulation (flow/capacity).

We will find a feasible flow (or determine if one does not exist) using a reduction to a maximum flow problem. We construct a graph G' as follows: Add two extra vertices s and t to graph G; connect the source s with every vertex v that has a negative demand; assign a capacity $-d(v)$ to each (s, v) edge; connect each vertex with a positive demand with the sink t; and assign a capacity $d(v)$ to each (v, t) edge. See figure 7.23 for an example.

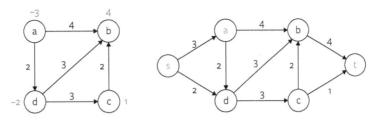

FIGURE 7.23   Left: Circulation with demands; right: A flow network.

The max flow in G' must saturate all the edges coming out of the source s; otherwise, there is no feasible solution.

Claim. *There is a feasible circulation with demands $d(v)$ in G if and only if the max-flow value in G' is $D = \sum_{v:\, d(v)>0} d(v)$.*

Proof. $\Rightarrow$) In graph G' we send $-d(v)$ units of flow along each edge from s, with the total flow $|f| = D$. Since there is a feasible circulation, that flow will reach the sink t, and moreover it is the maximum.

$\Leftarrow$) If the max-flow value in G' is D, then edges incident on s and t must be saturated. Remove those edges to get a feasible circulation. Figure 7.24 demonstrates a transformation from a flow network to a feasible circulation.

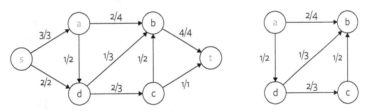

FIGURE 7.24   Left: A flow network; right: Feasible circulation.

7.5.2   Circulation with Lower Bounds

Now we impose restrictions on the edge capacity in a directed graph G. For every edge (u, v) we add a constraint $0 \le l(u, v) \le c(u, v)$, which is a lower bound to how much flow must be on this edge. By setting a lower bound $l(u, v) > 0$, we can force a particular edge to be used by flow.

We define a *circulation* with demands and lower bounds as a function f that assigns nonnegative real values to the edges of G and satisfies the following axioms:

1. Capacity constraint: $l(u, v) \leq f(u, v) \leq c(u, v)$, for each $u, v \in V$
2. Conservation constraint: $\sum_u f(u,v) - \sum_w f(v,w) = d(v)$, for each $v \in V$

We call a circulation *feasible* if it meets all these constraints. The question is if there exists feasible circulation. Figure 7.25 provides an example of a graph with demands on each vertex (in red) and capacity on each edge in the form $[l, c]$, meaning $l(u, v) \leq f(u, v) \leq c(u, v)$.

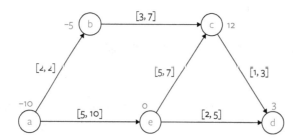

FIGURE 7.25   Circulation with demands and lower bounds.

We reduce this problem to the existence of a feasible circulation with demands. Let us start by pushing a flow f_0 on every edge with a value that is exactly equal to its lower bound $l(u, v)$. In the graph in figure 7.25, we push 2 units of flow on edge (a, b), 3 units on edge (b, c), 5 units on edges (a, e), and so on. A flow $f_0(u, v) = l(u, v)$ is a valid flow as far as capacities and lower bounds, but it might violate the conservation constraints. So, we need to compute

$$\sum_u f_0(u,v) - \sum_w f_0(v,w) = \sum_u l(u,v) - \sum_w l(v, w) = L(v)$$

for each vertex. If $L(v) = d(v)$, then flow f_0 satisfies the required demand. Otherwise, there is flow imbalance at vertex v. We fix this by transferring $L(v)$ to the vertex demand by setting a new demand $d'(v) = d(v) - L(v)$. In particular, for this graph, $L(e) = 5 - (5 + 2) = -2$ and $d(e) = 0 - (-2) = 2$. We have constructed a graph G' with new demands, $d'(v) = d(v) - L(v)$, and new capacities, $c'(u, v) = c(u, v) - l(u, v)$. See figure 7.26 for details.

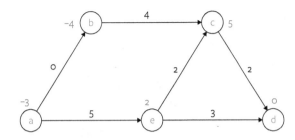

FIGURE 7.26 *G′ with new demands and capacities.*

Claim. *There is a feasible circulation in G if and only if there is a feasible circulation in G′.*

Proof. ⇒) Let f be a feasible circulation in G. Then by construction (we pushed an initial flow of the value $l(u, v)$ on each edge), $f'(u, v) = f(u, v) - l(u, v)$ is a feasible circulation in G'.

⇐) Let f' be a feasible circulation in G'. Construct a new flow, $f(u, v) = f'(u, v) + f_0(u, v)$. How do we find $f_0(u, v)$? Since we know the old $c(u, v)$ and new $c'(u, v)$ capacities on each edge, we compute $f_0(u, v) = l(u, v) = c(u, v) - c'(u, v)$. Next, we need to verify that f is a feasible circulation in G. First, we check the capacity constraints for circulation f:

$$l(u,v) \le f(u,v) \le c(u,v) \Leftrightarrow l(u,v) \le f'(u,v) + l(u,v) \le c'(u,v) + l(u,v) \Leftrightarrow 0 \le f'(u,v) \le c'(u,v).$$

Then we check the demand conditions for circulation f:

$$\sum_u f(u,v) - \sum_w f(v,w) = d(v) \Leftrightarrow$$

$$\sum_u (f'(u,v) + l(u,v)) - \sum_w (f'(v,w) + l(v,w)) = d'(v) + L(v) \Leftrightarrow$$

$$\sum_u l(u,v) - \sum_w l(v, w) + \sum_u f'(u,v) - \sum_w f'(v,w) = d'(v) + L(v) \Leftrightarrow$$

$$\sum_u f'(u,v) - \sum_w f'(v,w) = d'(v).$$

7.5.3 Circulation Problem Example

As an example of the circulation problem with demands and lower bounds, we consider the following problem:

Given the network (see figure 7.27) with the demand values on vertices and lower bounds on edge capacities, determine if there is a feasible circulation in this graph.

(a) *Turn the circulation with lower bounds problem into a circulation problem without lower bounds.*

(b) *Turn the circulation with demands problem into the maximum flow problem.*

(c) *Does a feasible circulation exist?*

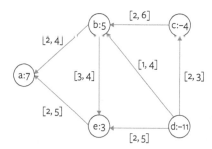

FIGURE 7.27   Circulation problem with demands and lower bounds.

Part (a): First, we check the necessary condition for a feasible circulation: The sum of demands must be equal to zero. Then we turn the circulation with lower bounds problem into a circulation problem without lower bounds. We push a flow with the value of the lower bound $l(u, v)$ on each edge and compute the flow excess $L(v) = f^{in}(v) - f^{out}(v)$ for each vertex v.

$$L(a) = (2 + 2) - 0 = 4,$$
$$L(b) = (2 + 1) - (2 + 3) = -2,$$
$$L(c) = 2 - 2 = 0,$$
$$L(d) = 0 - (1 + 2 + 2) = -5,$$
$$L(e) = (2 + 3) - 2 = 3.$$

Next, we recompute the demands $d'(v) = d(v) - L(v)$ to get

$$d'(a) = 7 - 4 = 3, d'(b) = 5 - (-2) = 7, d'(c) = -4 - 0 = -4,$$
$$d'(d) = -11 - (-5) = -6, d'(e) = 3 - 3 = 0.$$

We have reduced the original problem in a circulation problem without lower bounds. See figure 7.28.

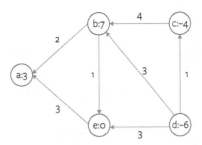

FIGURE 7.28 Circulation problem with no lower bounds.

Part (b): In order to reduce the circulation problem from part (a) into the max-flow network problem, we construct a new graph by adding two extra vertices, s and t. We connect the source s with vertices c and d by edges with capacities 4 and 6, respectively. We connect vertices a and b with the target t by edges with capacities 3 and 7, respectively. See figure 7.29 for the resulting graph.

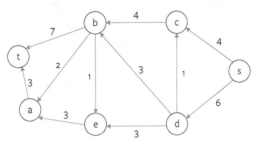

FIGURE 7.29 The max-flow network.

Part (c): Running the Ford–Fulkerson algorithm, we find that the max flow has value 10 and saturates all the edges coming out of the source s. Figure 7.30 is a feasible circulation to the original problem.

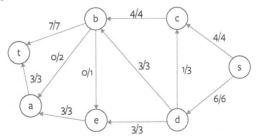

FIGURE 7.30 Network flow graph with flow/capacity on each edge.

7.6 Reduction to Circulation

As an example of using a reduction to circulation, consider the following problem:

> Consider *LAX*, a notoriously busy airport with many arriving passengers who want to get to their destinations as soon as possible. There is an available fleet of n Uber drivers to accommodate all passengers. However, there is a traffic regulation at the airport that limits the total number of Uber drivers at any given hour-long interval to $0 \leq k < n$ simultaneous drivers. Assume that there are p time intervals. Each driver provides a subset of the time intervals he or she can work at the airport, with the minimum requirement of a_j hour(s) per day and the maximum b_j hour(s) per day. Lastly, the total number of Uber drivers available per day must be at least m to maintain a minimum customer satisfaction and loyalty. Design an algorithm to determine if there is a valid way to schedule the Uber drivers with respect to these constraints.

We will reduce the Uber driver's problem to a circulation problem. First, we build a bipartite graph (see figure 7.31) having the drivers Ub_i on one side and hour-long time intervals I_j on the other side. We insert the edge between driver Ub_i and time interval I_j if the driver prefers to work at that hour. The capacity of this edge is 1. There could be many drivers willing to work at that hour, so having flow 0 on that edge is interpreted as a driver not covering that time interval.

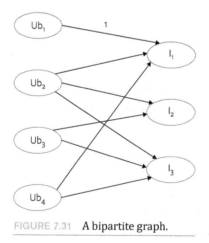

FIGURE 7.31   A bipartite graph.

Next, we add two new vertices x and y. Connect x to all Ub_i and all I_j to y. The edge (x, Ub_i) has lower bound a_i and upper bound b_i. The edge (I_j, y) has capacity k. Finally, we add the edge (y, x). The flow on this edge represents the total number of Uber drivers serving the airport. We set the lower bound on that edge to m. See figure 7.32 for the resulting graph H with $n = 4$ and $p = 3$.

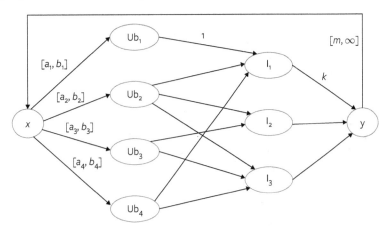

FIGURE 7.32   The Uber driver's problem as a circulation problem.

Claim. *There is a valid way to schedule the Uber drivers if and only if there is a feasible circulation in H.*

Proof. ⇒) Assume that there is a valid way to schedule at least m Uber drivers per day. We construct a flow in H as follows. If a driver Ub_i works during a time interval I_j, we create a flow of one unit on edge (Ub_i, I_j). A particular driver Ub_i may work during several time intervals. Therefore, we set the flow along the edge (s, Ub_i) to the number of time intervals that driver works. We set the flow along the edge (I_j, t) to the number of drivers who work during that time interval I_j. Finally, we set the flow on edge (t, s) to the total number of Uber drivers serving the airport. Thus, we have constructed a feasible circulation.

⇐) Consider a feasible circulation in H. For each edge (Ub_i, I_j) that carries one unit of flow, driver Ub_i works at hour I_j. Flow on the edge (s, Ub_i) represents the total number hours that driver works. By the flow conservation law, that number is between a_i and b_i. Similarly, the flow along the edge (I_j, t) cannot exceed k, implying that only at most k drivers can work at that hour I_j. ∎

If we want to know under what conditions a feasible circulation graph H exists, we need to turn the circulation problem into the max-flow network problem. We proceed as in section 7.4.2. See figure 7.33 for a newly constructed graph H' by removing the lower bounds and vertex demands. There we assume that $m - \sum_i a_i > 0$, so that vertex x is a supply. It follows that there is a feasible circulation in H if and only if the max-flow value in H' is m.

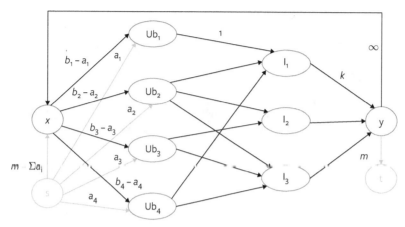

FIGURE 7.33   Graph H' for the max-flow problem.

REVIEW QUESTIONS

1. What is a flow?
2. What is a flow network?
3. What is an augmenting path?
4. What is the relationship between a flow value and a cut capacity?
5. Among all cuts, how do you distinguish a min-cut in the residual network?
6. How do you find a min-cut?
7. Is a min-cut unique?
8. How do you force the flow to use certain edges?
9. (**T/F**) A residual network is a flow network.
10. (**T/F**) The Ford–Fulkerson algorithm always terminates.
11. (**T/F**) The Ford–Fulkerson algorithm is a polynomial time algorithm.
12. (**T/F**) The Ford–Fulkerson algorithm is a greedy algorithm.
13. (**T/F**) The Edmonds-Karp 1 algorithm is a pseudo-polynomial time algorithm.
14. (**T/F**) The Edmonds-Karp 2 algorithm is a polynomial time algorithm.

15. **(T/F)** If all capacities in a flow network are integers, then every maximum flow in the network is such that the flow value on each edge is an integer.

16. **(T/F)** If we add the same positive number to the capacity of every directed edge, then the minimum cut (but not its value) remains unchanged.

17. **(T/F)** Given a max-flow value you can find a min-cut in $O(E)$.

18. **(T/F)** Given a min-cut value you can find a max-flow value in $O(E)$.

19. **(T/F)** Every flow is a circulation.

20. **(T/F)** There is a feasible circulation with demands $\{d_v\}$ if $\sum_v d_v = 0$.

EXERCISES

1. Given a flow network $N = (G = (V, E), s, t, c)$, where E might contain edges (u, v) and (v, u) in both directions for some pair of vertices u, v, we would like to use the Ford–Fulkerson algorithm to solve the flow problem on G, but G is not a flow network. Reduce this problem to the network flow problem.

2. Suppose we have a directed weighted graph $G = (V, E)$ with multiple sources $s_1, s_2, ..., s_n$ and multiple sinks $t_1, t_2, ..., t_m$. Reduce this problem to the network flow problem.

3. Given a flow network $N = (G = (V, E), s, t, c)$, find the maximum number of edge disjoint paths from s to t. A set of paths is edge disjoint if no two paths share an edge.

4. Given a flow network $N = (G = (V, E), s, t, c)$, find the maximum number of vertex disjoint paths from s to t. A set of paths is vertex disjoint if no two paths share a vertex.

5. Given a flow network $N = (G = (V, E), s, t, c)$, in which, in addition to having a capacity $c(u, v)$ for every edge, we also have a capacity $c(v)$ for every vertex. The flow coming to a vertex v cannot exceed the vertex capacity $c(v)$. Reduce this problem to the network flow problem.

6. You have successfully computed a maximum s-t flow for a network $G = (V, E)$ with positive integer edge capacities. Your manager now gives you another network G' that is identical to G except that the capacity of exactly one edge is decreased by one. You are also explicitly given the edge whose capacity was changed. Describe how you can compute a maximum flow for G' in linear time.

7. The vertex cover of an undirected graph $G = (V, E)$ is a subset of the vertices that touches every edge; that is, a subset $S \subset V$ such that for each edge $(u, v) \in E$, one or both of u, v are in S. Show that the problem of finding the minimum vertex cover in a bipartite graph reduces to the maximum flow problem.

8. A subset of edges is a matching if no two edges have a common vertex. A maximum matching is a matching with the largest possible number of edges. Our goal is to find the maximum matching in a bipartite graph. Show that the problem of finding the maximum matching in a bipartite graph reduces to the maximum flow problem.

9. There are n students in a class. We want to choose a subset of k students to join a committee. There has to be m_1 number of freshmen, m_2 number of sophomores, m_3 number of juniors, and m_4 number of seniors on the committee. Each student is from one of k departments, where $k = m_1 + m_2 + m_3 + m_4$. Exactly one student from each department has to be chosen for the committee. We are given a list of students, their home departments, and their class (freshman, sophomore, junior, or senior). Describe an efficient algorithm based on network flow techniques to select who should be on the committee such that these constraints are all satisfied.

10. Consider a set of mobile computing clients in a certain town who each need to be connected to one of several possible base stations. We'll suppose there are n clients, with the position of each client specified by its (x, y) coordinates in the plane. There are also k base stations; the position of each of these is specified by (x, y) coordinates as well. For each client, we wish to connect it to exactly one of the base stations. Our choice of connections is constrained in the following ways. There is a range parameter R, which means that a client can only be connected to a base station that is within distance R. There is also a load parameter L, which means that no more than L clients can be connected to any single base station. Given the positions of a set of clients and a set of base stations, as well as the range and load parameters, decide whether every client can be connected simultaneously to a base station.

11. The computer science department course structure is represented as a directed acyclic graph $G = (V, E)$ where the vertices correspond to courses and a directed edge (u, v) exists if and only if course u is a prerequisite for course v. By taking a course $w \in V$, you gain a benefit of p_w which could be a positive or negative number. Note, to take a course, you have to take all of its prerequisites. Design an efficient algorithm that picks a subset $S \subset V$ of courses such that the total benefit is maximized.

12. The edge connectivity of an undirected graph $G = (V, E)$ is the minimum number of edges that must be removed to disconnect the graph. For example, the edge connectivity of a tree is 1. Show how the edge connectivity of an undirected graph can be determined by running a maximum-flow algorithm.

13. There is a precious diamond that is on display in a museum at m disjoint time intervals. There are n security guards who can be deployed to protect the precious diamond. Each guard has a list of intervals for which he or she is available to be deployed. Each guard can be deployed to at most M time slots and has to be deployed to at least L time slots. Design an algorithm that decides if there is a deployment of guards to intervals such that each interval has either one or two guards deployed.

14. Your local police department has asked you to help set up the work shift schedule for the next month. There are n policemen on the staff and m days in the month. Each policeman gives a list of the days of the month that he or she is available to work. Let d_i denote the number of days that each policeman i is available to work. Then he or she should be scheduled to work at least $d_i/2$ of these days. Each day there must be exactly 2 policemen on duty. Design an algorithm that decides whether there exists a schedule that satisfies all of these requirements.

Chapter 8

Linear Programming

I N THIS CHAPTER WE WILL DESCRIBE a very general design technique called linear programming (LP). Like network flow and dynamic programming, it can be used to express a wide variety of linear optimization problems given certain constraints. We can use algorithms for linear programming to solve the shortest distance problem, the max-flow problem, and many other optimization problems. The latter especially includes problems of allocating resources and business supply-chain applications given limited resources and competing constraints.

The technique of linear programming was originally formulated by Russian economist L.V. Kantorovich in 1939. Later in 1975 he was awarded the Nobel prize in economics for contributions to the theory of optimum allocation of resources.

The word *programming* in linear programming is not used in the sense of computer programming as we understand it today. Its etymology is similar to dynamic programming (see chapter 6.) The world *linear* indicates the linear relationships between different variables.

8.1   Introduction: A Production Problem

Before we proceed with the theory, let us start with a motivating example.

> A jewelry company wishes to produce two types of rings: The first type will result in a profit of $100, and the second type in a profit of $120. To manufacture the first type of ring requires 2 rubies and 1 sapphire. The second type of ring requires 1 ruby and 3 sapphires. There are 200 rubies and 300 sapphires available. How many rings of each type should the company make in order to maximize its profit?

A linear programming problem consists of a linear objective function to be maximized or minimized, subject to certain constraints, in a form of linear equations or inequalities. First, we start with defining variables. Let $x \geq 0$ be the number of the first type rings and $y \geq 0$ be the number of the second type rings to be made. Then the total profit the company makes is given by $100\,x + 120\,y$. Therefore, the *objective* function for the problem is

$$\max_{x,\,y} 100x + 120y.$$

Next, we define constraints on x and y. The total amount of rubies is $2x + y$ is and must not exceed 200. The total amount of sapphire is $x + 3y$ and must not exceed 300. These lead to the following system of inequalities

$$2x + y \leq 200$$
$$x + 3y \leq 300$$
$$x, y \geq 0.$$

This is an example of a linear program: All our constraints are linear inequalities and the objective function is also linear.

We can solve our linear program by graphing the set of points in the plane that satisfies all the constraints and then finding the maximum of the objective function. A linear equation in x and y defines a line, and a linear inequality defines a *half space*, the region on one side of the line. Figure 8.1 represents a half space for inequality $2x + y \leq 200$.

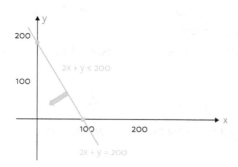

FIGURE 8.1   A half space for $2x + y \leq 200$.

Drawing other inequalities in the constraint set will give us a convex polygon S (see figure 8.2.) The set S (in blue) is the intersection of all four half spaces. Each point in S

is a candidate for the solution to our linear program and the whole set represents all *feasible* solutions.

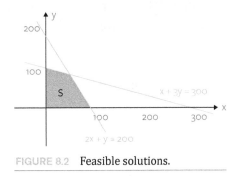

FIGURE 8.2 Feasible solutions.

We want to find a feasible point in S that maximized the objective function. For that, we draw an objective line $100\,x + 120\,y = p$, where p can take any real value and move it parallel to itself, up and to the right to get the larger and larger profit p (see figure 8.3). Ideally, we want to get as far as possible within the feasible region S and find the last point where the objective line intersects the feasible region. It is easy to see that the objective function always takes on its maximum value at a corner point of the feasible region. In our example that point is at the vertex (60, 80) and the objective function value is 15,600. The point(s) that optimizes the objective function of the linear program is called an *optimal* solution.

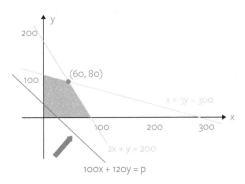

FIGURE 8.3 The arrow shows a direction of increasing profit.

Although it does not happen in our example, an entire polygon edge could be the optimal solution. This happens when an objective function line is parallel with one of the constraint lines. In this case a linear program has infinitely many optimal solutions.

8.1.1 Infeasibility and Unboundedness

Not all linear programs have solutions. In certain circumstances a linear program can be either infeasible or unbounded. Both situations are commonly due to shortcomings in the constraints formulation or to some wrong numbers in the data. Consider the following linear program:

$$\max_{x} x$$
$$x \leq 1$$
$$x \geq 2.$$

For this program, the constraint set S is empty. Since there is no assignment to the variables that satisfies all the constraints, the problem has no solution and is called *infeasible*.

Feasible sets may be bounded or unbounded. A problem is said to be *unbounded* if the constraints do not restrict the objective function and the optimal objective may be improved indefinitely. Here is an example:

$$\max_{x} x$$
$$x \geq 2.$$

If the feasible region is unbounded, the optimal objective value may or may not be finite. Consider the following unbounded linear program (depicted in figure 8.4), in which an objective function line is parallel to a constraint:

$$\max_{x, y} \ x - y$$
$$x - y \leq 1$$
$$x, y \geq 0.$$

The feasible region S is clearly unbounded, since any point $x = y$ belongs to it. On other hand, there is a finite solution to the problem, which occurs at the corner $x = 1$ and $y = 0$. Note that the solution is not unique; $x = 2, y = 1$ is another solution. Actually, the whole edge of the region S is a solution. We could make a unique solution by adding another constraint $x \leq 1$.

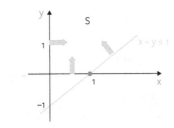

FIGURE 8.4   An unbounded linear program.

If a linear program is both feasible and bounded, then it has at least one finite optimal solution.

8.2   The Standard Maximum Problem

A linear program is the problem of optimizing a linear objective function in n variables, $x_1, ..., x_n$, subject to m linear inequalities. In standard inequality form, a linear program is written as

$$\max_{x_1, ..., x_n} (c_1 x_1 + ... + c_n x_n)$$

subject to

$$a_{11} x_1 + ... + a_{1n} x_n \le b_1$$
$$\vdots$$
$$a_{m1} x_1 + ... + a_{mn} x_n \le b_m$$

where each variable x_j satisfies the non-negativity constraint

$$x_1 \ge 0, ..., x_n \ge 0.$$

Most of the application problems do not automatically arise in standard form, though there is a variety of techniques to rephrase problems in standard form. All LP problems can be converted to standard form by the following techniques:

1.   A minimum problem can be changed to a maximum problem by multiplying the objective function by -1.

2. Constraints of the form $a\,x_j \geq b$ can be changed to $-a\,x_j \leq -b$.
3. An equality constraint $a\,x_j = b$ can be transformed into inequality form by replacing each equation by two inequalities, $a\,x_j \leq b$ and $-a\,x_j \leq -b$.
4. An unrestricted (free) variable x_j can be replaced by the difference of two variables, $x_j = u - v$, where $u \geq 0, v \geq 0$.
5. A variable constraint of the form $x_j \geq c$ can be transformed into $z_j \geq 0$ by replacing $x_j = z_j - c$.

The standard form is useful when we want to state theoretical results about linear programs without going through all special cases. From the application point of view, it's not necessary to convert a problem into standard form. The LP solver packages (like LINDO, CPLEX, Gurobi) carry out all necessary conversions.

A standard problem can be written in a matrix form if we introduce the following notations:

$$x = \begin{pmatrix} x_1 \\ x_2 \\ \dots \\ x_n \end{pmatrix}, c = \begin{pmatrix} c_1 \\ c_2 \\ \dots \\ c_n \end{pmatrix}, b = \begin{pmatrix} b_1 \\ b_2 \\ \dots \\ b_m \end{pmatrix}, A = \begin{pmatrix} a_{11} & \cdots & a_{1n} \\ \vdots & \ddots & \vdots \\ a_{m1} & \cdots & a_{mn} \end{pmatrix}.$$

By applying some basic linear algebra, this problem becomes

$$\max_x (c^T x)$$

subject to

$$A\,x \leq b$$
$$x \geq 0.$$

For example, in the production problem from chapter 8.1, we have

$$c = \begin{pmatrix} 100 \\ 120 \end{pmatrix}, \quad b = \begin{pmatrix} 200 \\ 300 \end{pmatrix}, \quad A = \begin{pmatrix} 2 & 1 \\ 1 & 3 \end{pmatrix}.$$

Every inequality of the form $a_{11}x_1 + \dots + a_{1n}x_n \leq b_1$ in the constraint set divides the space $\mathbb{R}^n$ into two regions, called *half spaces*, with the *hyperplane* $a_{11}x_1 + \dots + a_{1n}x_n = b_1$ being the boundary between them. An intersection of these half spaces forms a *polyhedron*, which is a convex set in n dimensions. A *polytope* is a bounded polyhedron. A cube and a tetrahedron are examples of polytopes. Corner points of a polytope are intersections of hyperplanes and called *extreme* points.

Theorem (Fundamental theorem). *The linear program either*

1. *has no optimal solution, in which case a feasible set is empty or unbounded; or,*

2. *has an optimal solution that must occur at one of the vertices of the polytope.*

In linear program we do not allow strict inequalities such as $a\,x < b$, since the solution is not guaranteed to exist at extreme points. Here is an example:

$$\max_x x$$
$$x < 2.$$

The maximum $x = 2$ does not lie in the feasible region.

An important consequence of this theorem is that an algorithm for solving a linear program only needs to examine all the extreme points of the polytope. How many vertices can there be? In a system with m constraints and n variables, that is the same as the number of ways to choose n linear independent rows from m rows, at most $\binom{n}{m}$. Thus, we have discovered an exponential time algorithm (in the worst case) for solving a linear program: Enumerate all vertices of the polytope, calculate the value of the objective function for each vertex, and take the maximum. This is the outline of an algorithm called the simplex algorithm, invented by G. Dantzig in 1947. The algorithm is very efficient in practice and runs in $O(n^2\,m)$ time in most cases, even with tens of thousands of variables and constraints (on modern computers).

The first polynomial time algorithm, the ellipsoid method, was discovered in 1979 by L. Khachian. The algorithm is terribly slow and not competitive with the simplex algorithm in practice, though it makes a theoretically powerful tool; for instance, it is used for combinatorial optimization problems. In 1984, N. Karmarkar described a faster polynomial time algorithm called the interior point method. However, the simplex algorithm remains the most popular method for solving linear programming problems.

8.3   A Few Applications

In this chapter we express problems that are familiar to us, for which we have developed efficient algorithms in the previous chapters, as linear programs. Though the linear programming approach is less efficient when using specialized algorithms, the main point here is to demonstrate how linear programming can be applied, and to illustrate its generality. Reducing a problem to linear programming may provide a quicker way (from a software engineering standpoint) to solve it, rather than to invent a custom algorithm for it.

8.3.1   The Shortest-Path Problem

In chapters 4.5 and 6.4 we have discussed the problem of finding the shortest directed path from s and t in a weighted directed graph $G = (V, E)$. In this section we will reduce the shortest path problem to linear programming. Recall the definition of a polynomial reduction from chapter 7.3.

We construct a linear program as follows. We define a variable $d(v)$ that denotes a distance from a source s to each vertex $v \in V$. Since the edge weights are allowed to be negative, each $d(v)$ is unrestricted in sign. For the source vertex, $d(s) = 0$. For every directed edge $(u, v) \in E$, we add the constraint $d(v) \leq d(u) + w(u, v)$, where $w(u, v)$ is the edge weight. This is illustrated in figure 8.5.

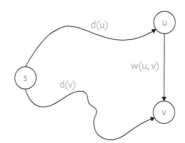

FIGURE 8.5   The relaxation constraint.

The relaxation constraint implies that $d(v) \leq \min_u (d(u) + w(u, v))$ is at most the shortest path distance from s to v. So, $d(v)$ is the largest value in the set $\{d(u) + w(u, v)\}$. It follows that in order to get the shortest distance to the target t, we need maximize $d(t)$. Another argument for why the objective function is to be maximized is that if we minimize $d(t)$ we will get a trivial solution. We do not require edge weights to be non-negative, but we have to watch out for negative weight cycles (see Exercise 8).

Here is the LP formulation for the single-source shortest-path problem, assuming no negative weight cycles:

$$\max d(t)$$

subject to

$$d(v) - d(u) \leq w(u,v), \text{ for every } (u,v) \in E$$

$$d(s) = 0, d(v) \text{ are unrestricted for every } v \in V \setminus \{s\}.$$

As an example, consider the directed weighted graph (figure 8.6). We need to calculate the shortest distance between s to t.

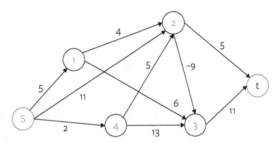

FIGURE 8.6   A shortest path problem.

This problem can be formulated as a following linear program:

$$\max d(t)$$

subject to

$$d(1) \leq d(s) + 5, d(2) \leq d(1) + 4, d(2) \leq d(4) + 5, d(2) \leq d(s) + 11, d(3) \leq d(4) + 13,$$
$$d(3) \leq d(2) - 9, d(3) \leq d(1) + 6, d(4) \leq d(s) + 2, d(t) \leq d(2) + 5, d(t) \leq d(3) + 11,$$

where

$$d(s) = 0, d(1) \geq -9, d(2) \geq -9, d(3) \geq -9, d(4) \geq -9, d(t) \geq -9.$$

Solving the above LP yields the correct results, $d(t) = 9$. However, the distance $d(1) = 3$ is underestimated. The optimal solution guarantees only the shortest distance from s to t. For other vertices, $d(v)$ may be an underestimate of the true distance. This could be easily fixed by changing the objective function (see Exercise 7).

What happens to the LP if there is no $s-t$ path in the given graph? Consider the following graph (figure 8.7) in which the vertex t is unreachable.

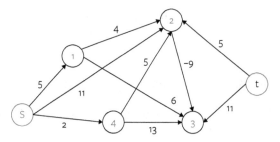

FIGURE 8.7 A graph with an unreachable vertex t.

A linear programming formulation is

$$\max d(t)$$

$$d(1) \leq d(s) +5 , d(2) \leq d(1) + 4, d(2) \leq d(4) + 5, d(2) \leq d(s) +11, d(3) \leq d(4) +13,$$

$$d(3) \leq d(2) - 9, d(3) \leq d(1) + 6, d(4) \leq d(s) +2, d(2) \leq d(t) + 5, d(3) \leq d(t) +11,$$

$$d(s) = 0, d(1) \geq -9, d(2) \geq -9, d(3) \geq -9, d(4) \geq -9, d(t) \geq -9.$$

As it turns out the linear program is unbounded. It follows from the last two constraints:

$$\begin{cases} d(2) \leq d(t) + 5 \\ d(3) \leq d(t) + 11 \end{cases}$$

We can readily calculate the shortest distances $d(2) = 7$ and $d(3) = -2$. The above inequalities therefore can be rewritten as

$$\begin{cases} d(t) \geq 2 \\ d(t) \geq -13 \end{cases}$$

which means that the maximum $d(t)$ cannot be reached.

8.3.2 The Max-Flow Problem

Recall the max-flow problem defined in chapter 7.1. Given a network $(G = (V, E), s, t, c)$ with a designated source s and sink t, and a nonnegative capacity $c(u, v)$ for each edge $(u, v) \in E$, we need to maximize the flow from s to t. The max-flow problem can be easily reduced to a linear program by following the definition of feasible flow. We introduce a variable $f(u, v)$ that denotes a flow across the edge $(u, v) \in E$. There are two types of constraints in a flow network:

1. Capacity constraint: $0 \leq f(u, v) \leq c(u, v)$, for each edge $(u, v) \in E$

2. Conservation constraint: $\sum_u f(u, v) = \sum_w f(v, w)$, for each $v \in V - \{s, t\}$

The objective function is to maximize a flow emanating from the source s (or descend to the sink t). The linear program has E variables and $2E + V - 2$ constraints. If we want to write this LP in the standard from, we need to change the equalities, $\sum_u f(u, v) = \sum_w f(v, w)$, into inequalities, $\sum_u - f(u, v) \leq \sum_w - f(v, w)$ and $\sum_u f(u, v) \leq \sum_w f(v, w)$.

As an example, consider the flow network in figure 8.8. We need to calculate the max-flow between s to t.

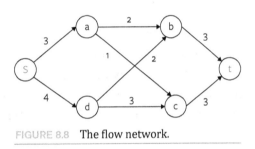

FIGURE 8.8   The flow network.

This problem can be formulated as a following linear program:

$$\max f(s, a) + f(s, d)$$

subject to

$$0 \leq f(s, a) \leq 3, 0 \leq f(s, d) \leq 4, 0 \leq f(a, b) \leq 2, 0 \leq f(a, c) \leq 1,$$
$$0 \leq f(b, t) \leq 3, 0 \leq f(c, t) \leq 3, 0 \leq f(d, c) \leq 3, 0 \leq f(d, b) \leq 2,$$
$$f(s, a) = f(a, b) + f(a, c), f(a, b) + f(d, b) = f(b, t),$$
$$f(a, c) + f(d, c) = f(c, t), f(s, d) = f(d, b) + f(d, c).$$

8.3.3   The Knapsack Problem

Recall the knapsack problem from chapter 6.1. You are given a set of n unique items, with weights $w_1, ..., w_n$ and values $v_1, ..., v_n$, where the weights and values are all integers. The problem is to find a subset of the most valuable items such that their total weight does not exceed W. We assume that all items are unbreakable. We formalize the problem by introducing an indicator variable x_k for each item $k = 1, 2, ..., n$:

$$x_k = \begin{cases} 1, \text{ if item } k \text{ is selected} \\ 0, \text{ otherwise.} \end{cases}$$

Then, we write the 0-1 Knapsack problem as follows

$$\max \sum_{k=1}^{n} v_k x_k$$

$$\sum_{k=1}^{n} w_k x_k \leq W$$

Notice that since the variables $x_k \in \{0, 1\}$ are integers, we do not have an ordinary linear program. This is an integer linear programming (ILP) problem that cannot be solved by the Simplex method. It's even worse: There is no polynomial algorithm that solves this problem. On the other hand, there is no proof that such an algorithm does not exists. We will prove in chapter 9 that ILP is a NP-hard problem.

8.4 The Dual Linear Program

Generally, the duality principle allows us to prove that a solution to an optimization problem is optimal. In chapter 7.2.3 we have used duality to prove the maximum flow. In this chapter we will describe how to formulate a dual linear program in which we minimize an objective function. We call the original linear program the primal. The dual of a dual linear program is the original primal linear program.

Given a primal linear program in standard maximum form

$$\max_x (c^T x)$$
$$\text{subject to} \quad Ax \leq b$$
$$x \geq 0$$

we define the dual as standard minimum problem:

$$\min_{y}(b^T y)$$

$$\text{subject to} \quad A^T y \geq c$$

$$y \geq 0$$

As an example, consider the production problem from section 8.1:

$$\max_{x,\,y} 100x + 120y$$

$$2x + y \leq 200$$

$$x + 3y \leq 300$$

$$x, y \geq 0$$

Here, the variables x and y represent the number of the first and second types of rings correspondingly. The dual to the previous linear program is in the variables u and v, which represent the *shadow* prices.

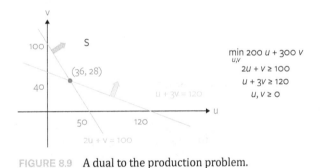

FIGURE 8.9   A dual to the production problem.

A shadow price is the value per unit of a resource, which in our case is rubies and sapphires. Figure 8.9 shows that the objective function takes its minimum value at a corner point of the feasible region, at the vertex with coordinates $(36, 28)$. These numbers are the minimal prices at which we are willing to sell each ruby and sapphire. If we compute the value of resources based on shadow prices, we get our optimal profit: $200 \times 36 + 300 \times 28 = 15,600$. The objective function value is the same as in the primal problem. This is not a coincidence but reflects a fundamental property of primal and dual programs.

The relation between a standard problem and its dual is given in the following theorems and corollaries.

Theorem 1. (The weak duality) *Let P and D be primal and dual LP correspondingly. If x is a feasible solution for P and y is a feasible solution for its dual D, then* $c^T x \leq b^T y$.

Proof. Since y is feasible solution, we have

$$c^T x = x^T c \leq x^T (A^T y) = (A x)^T y \leq b^T y.$$

The first inequality follows from the fact that y is feasible solution; the second inequality follows since x is feasible solution. ∎

The theorem says that the optimum of the dual is an upper bound to the optimum of the primal. The difference between them, $c^T x - b^T y$, is called a *duality gap*.

Corollary 1. *If a standard problem and its dual are both feasible, then both are feasible bounded.*

Proof. Since the dual is an upper bound to the optimum of the primal, then the primal is bounded. If the primal is a lower bound to the optimum of the dual, then the dual is bounded. ∎

Corollary 2. *If one problem has an unbounded solution, then the dual of that problem is infeasible.*

Proof. Suppose that the dual is feasible. Then, the dual would provide an upper bound on the primal. This contradicts the fact that the primal problem is unbounded. The argument for the dual is analogous. ∎

Theorem 2. (The strong duality) *Let P and D be primal and dual LP correspondingly. If x is a feasible solution for P and y is a feasible solution for its dual D, then* $c^T x = b^T y$.

The proof of this theorem is beyond the scope of this book.

Table 8.1 demonstrates all possible relations between the primal P and the dual D. In the table we use the following notations: F.B. (feasible bounded), F.U. (feasible unbounded), I. (infeasible). The NO in the table shows the impossibilities that follow either from Corollary 1 or 2. The YES in the table means the possibility, and we provide a corresponding example.

TABLE 8.1   Relations between the primal and the dual

P/D	F.B.	F.U.	I.
F.B.	YES (1)	NO	NO
F.U.	NO	NO	YES (2)
I.	NO	YES (3)	YES (4)

Example 1. (P and D are F.B.)

$$\max(x_1 + 4x_2)$$
$$x_1 + x_2 \leq 3$$
$$x_1 - x_2 \leq 2$$
$$x_1, x_2 \geq 0$$

$$\min(3y_1 + 2y_2)$$
$$y_1 + y_2 \geq 1$$
$$y_1 - y_2 \geq 4$$
$$y_1, y_2 \geq 0$$

Example 2. (P is F.U. and D is I.)

$$\max(x_1 + 4x_2)$$
$$x_1 - x_2 \leq 3$$
$$x_1 - x_2 \leq 2$$
$$x_1, x_2 \geq 0$$

$$\min(3y_1 + 2y_2)$$
$$y_1 + y_2 \geq 1$$
$$-y_1 - y_2 \geq 4$$
$$y_1, y_2 \geq 0$$

Example 3. (P is I. and D is F.U.)

$$\max(x_1 + 4x_2)$$
$$x_1 + x_2 \leq -3$$
$$x_1 - x_2 \leq 2$$
$$x_1, x_2 \geq 0$$

$$\min(-3y_1 + 2y_2)$$
$$y_1 + y_2 \geq 1$$
$$y_1 - y_2 \geq 4$$
$$y_1, y_2 \geq 0$$

Example 4. (*P* and *D* are I.)

$$\max(x_1 + 4x_2) \qquad\qquad \min(-3y_1 + 2y_2)$$
$$-x_1 + x_2 \le -3 \qquad\qquad -y_1 + y_2 \ge 1$$
$$x_1 - x_2 \le 2 \qquad\qquad y_1 - y_2 \ge 4$$
$$x_1, x_2 \ge 0 \qquad\qquad y_1, y_2 \ge 0$$

Theorem 2 states that if the primal and dual problems have optimal solutions, then the optimal objective function values must be equal. But it does not mean that a duality gap of the linear program is always zero. It is possible for both the primal and dual problems to be infeasible. In this case the duality gap is infinity (see Exercise 14).

REVIEW QUESTIONS

1. What is linear programming?
2. What is an objective function?
3. What are the nonnegativity constraints?
4. What is an optimal solution?
5. What is a feasible solution?
6. (**T/F**) Every LP has an optimal solution.
7. (**T/F**) If an LP has an optimal solution it occurs at an extreme point.
8. (**T/F**) If an LP is feasible and bounded, then it must have an optimal solution.
9. (**T/F**) An LP allows strict inequalities in the constraints.
10. (**T/F**) An LP for which the feasible region is unbounded has the finite optimal solution.
11. (**T/F**) The weak duality theorem does not always hold for an integer linear program.
12. (**T/F**) An LP must be infeasible if its dual problem is unbounded.
13. (**T/F**) Both the primal and the dual can be infeasible.
14. (**T/F**) There is no duality gap in linear programming.

1. A furniture company produces two types of chairs. The first type takes 10 hours to make and uses 2 square yards of fabric and 20 pounds of padding. The second type takes 70 hours to make and uses 3 square yards of fabric and 10 pounds of padding. The profit of the first type is $2 per chair, and the profit of the second type is $5 per chair. The resources available for production for both chairs are 490 hours of labor, 32 yards of fabric, and 240 pounds of padding. How many chairs of each type should the company make in order to maximize its profit?

2. A cargo plane can carry a maximum weight of 100 tons and a maximum volume of 60 cubic meters. There are three materials to be transported, and the cargo company may choose to carry any amount of each, up to the maximum available limits provided in the table below

	Density	Volume	Price
Material 1	2 tons/m³	40 m³	$1,000 per m³
Material 2	1 tons/m³	30 m³	$2,000 per m³
Material 3	3 tons/m³	20 m³	$12,000 per m³

Write a linear program that optimizes revenue given the constraints.

3. A furniture company produces three types of couches. The first type uses 1 foot of framing wood and 3 feet of cabinet wood. The second type uses 2 feet of framing wood and 2 feet of cabinet wood. The third type uses 2 feet of framing wood and 1 foot of cabinet wood. The profit of the three types of couches is $10, $8, and $5, respectively. The factory produces 500 couches each month of the first type, 300 of the second type, and 200 of the third type. However, this month there is a shortage of cabinet wood to only 600 feet, but the supply of framing wood is increased by 100 feet. How should the production of the three types of couches be adjusted to minimize the decrease in profit?

4. You have $1,000 to invest. There are three types of investments. The first type is every dollar invested yields $0.10 a year from now and $1.30 three years from now. The second type is every dollar invested yields $0.20 a year from now and $1.10 two years from now. The third type is every dollar invested a year from now yields $1.50 three years from now. The most that can be invested into a single investment is $500. During each year all leftover cash is placed into money markets that yield 6% per year. Write a linear program to maximize your investment in three years from now.

5. The Canine Products company has two dogfood products, Frisky Pup and Husky Hounds, that are made from a blend of two raw materials, cereal and meat. One pound of cereal and 1.5 pounds of meat are needed to make a package of Frisky Pup, and it sells for $7 a package. Two pounds of cereal and 1 pound of meat are needed to make a package of Husky Hound, and it sells for $6 a package. Raw cereal costs $1 per pound and raw meat costs $2 per pound. It also costs $1.40 to package the Frisky Pup and $.60 to package the Husky Hound. A total of 240,000 pounds of cereal and 180,000 pounds of meat are available per month. The production bottleneck is that the factory can only package 110,000 bags of Frisky Pup per month. Write a linear program to maximize profit.

6. Rewrite the following linear programs in the standard maximum form:

   a. Maximize $2x + 3y$

   subject to $5x - 6y \geq 7$

   $7x + 8y \leq 9$

   $x \geq 0, y \geq 2$

   b. Maximize $2x + 3y$

   subject to $5x - 6y \geq 7$

   $7x + 8y = 9$

   $x \geq 0$

   c. Minimize $5x - 2y + 9z$

   subject to $3x + y + 4z = 8$

   $2x + 7y - 6z \leq 4$

   $x \leq 0, z \geq 1$

7. Modify the linear program in section 8.3.1 to find the shortest distance from the source s to all other vertices.

8. What happens to the LP in section 8.3.1 if a given graph has negative weight cycles?

9. The all-pairs shortest-paths problem is to find a shortest path between any pair of vertices, u to v. Formulate the all-pairs shortest-paths problem as a linear program.

10. Given a bipartite graph, $G = (V, E)$, a subset of edges is a matching if no two edges have a common vertex. A maximum matching is a matching with the largest possible number of edges. Our goal is to find the maximum matching in a bipartite graph G. Write a linear program that solves the maximum-matching problem.

11. There are n people and n jobs. You are given a cost matrix, where each element $C(i, j)$ represents the cost of assigning person i to do job j. You need to assign all the jobs in such a way that each person performs only one job and each job is assigned to only one person. Write a linear program that minimizes the total cost of the assignment.

12. Given an infinite supply of bins, each of which can hold the maximum weight of 1, and there are also n objects, each of which has a weight $w_i \leq 1$, your goal is to place all the objects into bins in such a way that the total number of used bins is minimized. Formulate the problem as an integer linear programming problem.

13. Write the duals to the following linear programs:

    a. Maximize $x_1 + x_2 + 2x_3$

       subject to $x_1 + 2x_3 \leq 3$

                     $-x_1 + 3x_3 \leq 2$

                     $2x_1 + x_2 + x_3 \leq 1$

                     $x_1, x_2, x_3 \geq 0$

    b. Maximize $3x_1 - 2x_2 + x_3$

       subject to $x_1 - x_2 + x_3 \leq 4$

                     $3x_1 + x_2 + 2x_3 \leq 6$

                     $-x_1 + 2x_3 = 3$

                     $x_1 + x_2 + x_3 \leq 8$

                     $x_1, x_2, x_3 \geq 0$

    c. Minimize $3y_1 - 2y_2 + 5y_3$

       subject to $-y_2 + 2y_3 \geq 1$

                     $y_1 + y_3 \geq 1$

                     $2y_1 - 3y_2 + 7x_3 \geq 5$

                     $y_1, y_2, y_3 \geq 0$

14. Create an example of a linear program showing that the strong duality theorem does not always hold for an integer linear program.

15. Create an example of a linear program showing that the primal and the dual can be both infeasible.

Chapter 9

NP Completeness

I N PREVIOUS CHAPTERS WE HAVE SEEN different algorithms that run in worst-case polynomial time. We say that those algorithms are efficient. At the same time, we have seen problems that cannot be computed in polynomial time. That raises two questions: What is computable? and What is efficiently computable? These are the fundamental questions of computer science. To answer these questions, we have to introduce an abstract model of computation—the Turing machine. Turing machines provide a precise, formal definition of what it means to be computable. In this chapter we will consider a class of hard problems for which it is unknown if they can be solved in polynomial time. At the heart of these is the most famous unsolved problem in computer science: $\mathcal{P}$ versus $\mathcal{NP}$.

9.1   A Brief Introduction to the Turing Machines

In the 1900 International Congress of Mathematicians, D. Hilbert presented a list of 23 challenging (unsolved) problems in mathematics. One of them (known today as Hilbert's 10th problem) was formulated (my rephrasing) as follows:

> Given a multivariate polynomial with integer coefficients, devise a process according to which it can be determined, in a finite number of operations, whether it has an integer root.

In modern terminology, Hilbert was asking for an algorithm to decide whether a Diophantine equation has a solution in integers. This problem sparked the great interest in the research community. For many years people have tried to devise such an algorithm without success. Eventually, they began to think that it could not be done at all, so they started to search for proof that there is no such algorithm at all. Only in the mid-1930s did Alonzo Church and Alan Turing show that some problems have no algorithmic solution. In other words, they are unsolvable. Turing's proof introduced the notion of computation by machine, nowadays called the Turing machine. The machine precisely defines the meaning of an algorithm. An algorithm is a Turing machine in the sense that if an algorithm exists, then a Turing machine can run it. We say that a problem is *computable* if there is an algorithm for solving it in a finite number of steps. Therefore, an algorithm must always halt.

The *Turing machine* is a computing device, consisting of a head with a tape of unbounded length passing through it—a tape divided into cells. Each cell contains one symbol. The machine can perform only the following types of operations—read, write, move left, move right, change state, and halt. Based on the symbol it is currently reading, and its current state, the Turing machine either writes a new symbol in that location, moves to a new state, or stays in place. Once the computation is completed, the machine will come to a halt state. Figure 9.1 shows an example of a Turing machine that takes a binary string and appends 0 to its left side.

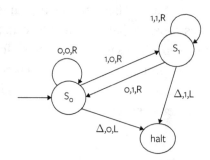

FIGURE 9.1   An Example of a Turing machine.

The states are represented by vertices and the transitions are represented by edges. Each transition has a triple of the form read, write, and direction. For example, (0,1,R) means if reading a 0, then write a 1 and move the head right. Δ denotes an empty cell. Computation starts at state S_0. If, for example, the first character of the input is a 1, we output a 0, move the head to the next character, and transition to state S_1. A computation may consist of millions of transitions. The Turing machines we have described here are deterministic: For every state there must be exactly one transition.

Despite its simplicity, the Turing machine is capable of computing anything that the modern computer can compute. According to the Church-Turing thesis (conjecture), everything that can be computed can be computed by a Turing machine. That is not a theorem; it has not been and cannot be proven. Also, as of today no counterexample has yet been constructed.

9.2   Computational Intractability

With the Turing machine we are ready to define the runtime complexity and complexity classes. The runtime complexity is the function $f\colon \mathbb{N} \to \mathbb{N}$ such that $f(n)$ is the maximum number of steps (transitions) that the Turing machine uses on any input of length n.

Definition. *A fundamental complexity class $\mathcal{P}$ (or PTIME) is a class of decision problems that can be solved by a deterministic Turing machine in polynomial time.*

A fundamental complexity class EXPTIME is a class of decision problems that can be solved by a deterministic Turing machine in $O(2^{p(n)})$ time, where $p(n)$ is a polynomial. By a *decision* problem we mean a problem that can be formulated as a "yes-no" question. Considering decision problems only does not reduce the scope of all problems, since every computational problem is equivalent to a decision problem. For instance, any optimization problem can be converted into a decision problem (see Exercise 1.)

A decision problem is *decidable* if it can be solved by a Turing machine that always halts; otherwise, it is called *undecidable.* An undecidable problem cannot be solved by any Turing machine. The most famous example of an unsolvable problem is the *halting problem.* That is the problem of whether a given Turing machine will terminate on a given input or instead it will run forever.

Theorem (A. Turing, 1936). *The halting problem is undecidable (unsolvable).*

Proof. We will prove it by a self-referencing contradiction as in the famous liar's paradox, the one about saying, "I am lying." If that statement is true, then it's not true. But if the statement is not true, then it is true.

Let $P(x)$ denote the output that arises from running program P on input x, assuming that P eventually halts. Then $P(P)$ means the output obtained from running program P on the text of its own source code. Let K be the set of all programs P such that $P(P)$ halts:

$$K = \{\text{program } P \mid P(P) \text{ halts}\}.$$

Clearly a set K is not empty; a Java program could be an element of that set. Next, we define a program HALT as follows:

$$\text{HALT(P)} = \begin{cases} \text{yes,} & \text{if } P \in K \text{, so } P(P) \text{ halts.} \\ \text{no, if } P \notin K \text{, so } P(P) \text{ doesn't halt.} \end{cases}$$

Let us assume that such program HALT does exist. Finally, we define a new program CONFUSE that calls HALT as a subroutine:

```
bool CONFUSE(P) {
  if (HALT(P) == True)
       then loop forever;
  else return True;
}
```

Does CONFUSE(CONFUSE) halt? Consider two cases:

1. Assume CONFUSE(CONFUSE) does halt.

    Then, by the definition of program HALT, we have that HALT(CONFUSE) is true. And by the definition of program CONFUSE, we have that CONFUSE(CONFUSE) loops forever.

2. Assume CONFUSE(CONFUSE) does not halt.

    Then, by the definition of program HALT, we have that HALT(CONFUSE) is false. And by the definition of CONFUSE, we have that CONFUSE(CONFUSE) returns true.

This is a contradiction. We have assumed that HALT exists; therefore, such a program HALT cannot exist. ∎

Why is the halting problem so important? There are two reasons. First, a lot of practical problems are the halting problem in disguise. For example, there is no algorithm that can reliably detect all software viruses. Second, if the halting problem could be solved, many other problems could be decided. For example, the famous Goldbach's conjecture could be decided. This conjecture states that every even integer greater than 2 can be expressed as the sum of two primes. We can write a program that runs until it finds the first counterexample to Goldbach's conjecture. If the halting problem was decidable then Goldbach's conjecture would be true if this program never halted and would be false if it did halt.

In his original 1936 paper, Turing also defined an extension of his deterministic machine that is known today as *nondeterministic* Turing machines. However, the concept of nondeterminism did not get much interest until works by M. Rabin and D. Scott in the early 1960s. Formally, a nondeterministic Turing machine has all the components of a standard deterministic Turing machine, except that at every state there is a set of possible transitions, any of which can be chosen by the machine. Therefore, a nondeterministic machine specifies a computation rooted tree. In this tree, a path from the root to a leaf is a computation. In a deterministic machine, the computation tree is just a single path. The power of a nondeterministic Turing machine is that it does computations in parallel. Using this machine, we can define new computational classes.

Definition. *A fundamental complexity class $\mathcal{NP}$ is a class of decision problems that can be solved by a nondeterministic Turing machine in polynomial time.*

For example, consider the problem of coloring the vertices of a graph with $k > 2$ colors so that no two adjacent vertices have the same color belongs to the $\mathcal{NP}$ class. A nondeterministic algorithm can simply guess an assignment of colors and then check in polynomial time if all pairs of adjacent vertices have distinct colors.

There is another view of the $\mathcal{NP}$ class that uses an alternative verifier-based definition.

Definition. *A fundamental complexity class $\mathcal{NP}$ is a class of decision problems where each provides a certificate that can be verified by a deterministic Turing machine in polynomial time.*

Consider the Hamiltonian path problem (see chapter 1.3.6). Assume we were given a sequence of vertices. We could verify in polynomial time whether they form a Hamiltonian path by visiting all vertices in the sequence. $\mathcal{NP}$ problems can be viewed as finding a needle in a haystack: It is hard to find it but it's always easy to verify once the needle is found.

Unfortunately, a mighty nondeterministic machine is a pure abstraction since no physical computer (even a quantum computer) can support unlimited parallelism. It is easy to see (running a breadth-first search) that a deterministic machine can recompute the entire computational tree of a nondeterministic machine. We can state that if a problem can be solved by a nondeterministic Turing machine, then it can be solved by a deterministic one. The difficulty is that such simulation between machines takes exponential time. But can we do it efficiently (i.e., in polynomial time)? The famous $\mathcal{P}$ versus $\mathcal{NP}$ conjecture would answer this question: We cannot hope to simulate nondeterministic

Turing machines in polynomial time. Therefore, we believe that these two classes are not equal since researchers have devoted an enormous amount of time trying to find polynomial time algorithms for some $\mathcal{NP}$ problems without success.

Next, we describe two more complexity classes: the $\mathcal{NP}$-hard and $\mathcal{NP}$-complete. For that we need to recall the definition of polynomial reduction (see chapter 7.3)

Definition. *A polynomial-time reduction of a decision problem Y to a decision problem X (we write it as $Y \leq_p X$) is a map $f: Y \rightarrow X$ such that*

1. *f is a polynomial time computable, and*
2. *$\forall y \in Y$ is yes-instance if and only if $f(y) \in X$ is yes-instance.*

In the previous chapters we use reductions to solve problems. A reduction $Y \leq_p X$ means that if we have an algorithm for problem X, we can use it to solve problem Y following these steps:

- Reduce an input of Y into an input of X

- Solve X

- Reduce the solution back to Y

In particular, if we can solve X in polynomial time, then we can solve Y in polynomial time.

In this chapter we use reductions to show that we cannot solve some problems. The contrapositive of the previous statement is, "If we cannot solve Y in polynomial time, then we cannot solve X in polynomial time." Therefore, the second meaning of $Y \leq_p X$ is that knowing that problem Y is hard (it has no an efficient algorithm), we prove that X is at least as hard as Y.

An example is Independent Set $\leq_p$ Vertex Cover.

Recall the definitions of an independent set and a vertex cover from chapter 1.3.6. Given a graph $G = (V, E)$. A set of vertices C is a vertex cover if every edge in E has at least one endpoint in C. A set of vertices I is an independent set if no two vertices of I are connected by an edge of E. We define decision problems as follows:

Vertex cover problem: Given a graph G and integer $k > 0$, decide whether there is a vertex cover of size k.

Independent set problem: Given a graph G and integer $k > 0$, decide whether there is an independent set of size k.

The proof of reduction from an independent set problem to a vertex cover problem is based on the fact that a graph $G = (V, E)$ has an independent set of size $\geq k$ if and only if G has a vertex cover of size $\leq V - k$.

Definition. *A problem X is in $\mathcal{NP}$-hard is for any $Y \in \mathcal{NP}$ it holds that $Y \leq_p X$.*

Definition. *A decision problem X is in $\mathcal{NP}$-complete if $X \in \mathcal{NP}$ and $X \in \mathcal{NP}$-Hard.*

An $\mathcal{NP}$-hard problem does not necessarily belong to the $\mathcal{NP}$ class. The halting problem is an example. Also, not all $\mathcal{NP}$-hard problems are decision problems; some of them are optimization problems. We already know that these two kinds of problems are essentially equivalent. However, they belong to two different complexity classes.

These are two of the most important and interesting classes of problems. If an $\mathcal{NP}$-hard (or $\mathcal{NP}$-complete) problem can be solved in polynomial time, then all $\mathcal{NP}$ (and $\mathcal{NP}$-complete) problems will be solved in polynomial time. Therefore, if one solves such a problem, it would follow that $\mathcal{P} = \mathcal{NP}$.

The diagram in figure 9.2 shows the graphical relationships between different complexity classes, assuming $\mathcal{P} \neq \mathcal{NP}$.

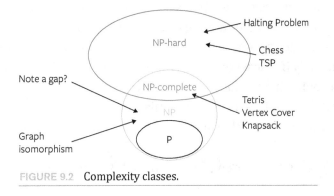

FIGURE 9.2   Complexity classes.

9.3   NP-Complete Problems

In this section we will take a look at a specific class of problems for which no efficient algorithms are known. These are $\mathcal{NP}$-complete problems. As we learned in the previous section, if a polynomial time algorithm were to be found for any one of these problems, then we could derive polynomial time algorithms for all of the problems in this class. There are hundreds of $\mathcal{NP}$-complete problems that have been identified.[1]

---

[1]"NP-Complete Problems," Wikipedia, https://en.wikipedia.org/wiki/List_of_NP-complete_problems

In this chapter we will consider only a few of them. Our first example is the satisfiability problem, which we will call SAT for short.

9.3.1   SAT Problem

Given a logical formula consisting of Boolean variables and operators AND (conjunction, $\wedge$), OR (disjunction, $\vee$), NOT (negation, $\neg$), we say that a formula is in conjunctive normal form (CNF) if it is a conjunction of clauses, where each clause is a disjunction of literals. A literal is a Boolean variable or its negation. For example,

$$(x_1 \vee x_2) \wedge (x_1 \vee \neg x_2 \vee x_4 \vee x_5) \wedge (x_2 \vee x_3 \vee \neg x_5)$$

A formula is in disjunctive normal form (DNF) if it is a disjunction of clauses, where each clause is a conjunction of literals:

$$(x_1 \wedge x_3 \wedge \neg x_5) \vee (x_1 \wedge \neg x_2 \wedge x_5) \vee (\wedge x_3 \wedge \neg x_4)$$

A formula is said to be *satisfiable* if it can be made TRUE by assigning appropriate logical values (TRUE, FALSE) to its variables. Therefore, SAT is the problem of determining if there exists an assignment that satisfies a given formula. We would like to find an algorithm whose worst-case running time is polynomial in the number of variables. It is easy to see that such an algorithm exists for DNF satisfiability. Any DNF formula is satisfiable if and only if at least one of its clauses is satisfiable. A conjunctive clause is satisfiable if and only if it does not contain both a literal and its negation. However, CNF satisfiability is $\mathcal{NP}$-complete, so no polynomial time algorithm has been found yet. At the same time, it is not proven that such an algorithm does not exists.

Theorem 1. (Cook-Levin theorem, 1971) *CNF-SAT is $\mathcal{NP}$-complete.*

We won't prove the theorem here since it's beyond the scope of the book. This result seems paradoxical, because using De Morgan's laws

$$\neg(x \wedge y) = \neg x \vee \neg y$$
$$\neg(x \vee y) = \neg x \wedge \neg y$$

we can convert any CNF formula into an equivalent DNF formula. The catch is that this conversion may require an exponential number of variables (or clauses) in the worst-case scenario.

Once we have one $\mathcal{NP}$-complete problem, the task of showing other problems to be $\mathcal{NP}$-complete becomes much easier, since we can use a polynomial reduction between two problems. To show that a problem X is $\mathcal{NP}$-complete, we will follow these three steps:

1. Show that X is in $\mathcal{NP}$
2. Pick a problem Y, known to be an $\mathcal{NP}$-complete
3. Show that X is in $\mathcal{NP}$-hard, namely prove $Y \leq_p X$

This is the technique that we will use for all subsequent NP-completeness (and $\mathcal{NP}$-hardness) proofs. In order to illustrate this technique, we consider a special case of Boolean satisfiability. We say that a CNF formula is k-CNF if no clause contains more than k literals. Therefore, k-SAT is the problem of determining satisfiability for a given k-CNF formula. Next, we will show that 3-SAT is $\mathcal{NP}$-complete.

Theorem 2. *3-SAT is $\mathcal{NP}$-complete.*

Proof. The fact that 3-SAT is in $\mathcal{NP}$ follows immediately from the observation that if we have a truth assignment, we can substitute it into a given 3-CNF and then evaluate the expression in polynomial time. Another way to prove that 3-SAT is in $\mathcal{NP}$ is to non-deterministically guess values for all the variables and then evaluate the formula. This can be done in nondeterministic polynomial time.

To prove $\mathcal{NP}$-hardness, we reduce CNF-SAT to 3-SAT. Since instances of CNF-SAT are already in CNF, we only need to ensure the number of literals in each clause. We will do this by breaking up clauses that are too long into clauses containing only 3 literals. For clauses with three literals or less, we do nothing. Consider a clause with four literals $(a \vee b \vee c \vee d)$ and let us break it into two clauses of 3 literals. The first obvious try is as follows

$$(a \vee b \vee c \vee d) \rightarrow (a \vee b \vee c) \wedge (b \vee c \vee d).$$

Unfortunately, it does not work, since not every assignment that satisfies the left-hand side of the expression will satisfy the right-hand side. Indeed, $a = \text{T}, b = c = d = \text{F}$ is an example of such an assignment. We learn from this example that we need leverage, namely a free variable that does not belong to a given SAT. Let us introduce a new variable, x, and replace $(a \vee b \vee c \vee d)$ with the following conjunction of clauses:

$$(a \vee b \vee c \vee d) \rightarrow (a \vee b \vee x) \wedge (\neg x \vee c \vee d).$$

Note that this statement is not a logical equivalence, since there are different variables on both sides of the statement.

If there is a truth assignment that satisfies the left-hand side, then at least one of its literals must be true. Let a = T. Then, to satisfy the right-hand side, we need to satisfy $(\neg x \vee c \vee d)$ for any c and d. We do this by setting the extra variable $x = F$.

We now claim that any assignment that satisfies the new clauses will also satisfy $(a \vee b \vee c \vee d)$. We prove this by contradiction. Suppose that $(a \vee b \vee c \vee d)$ is not satisfied (i.e., $a = b = c = d = F$). In order for the first new clause $(a \vee b \vee x)$ to be satisfied, the variable x must be true. Then the second clause $(\neg x \vee c \vee d)$ is not satisfied—a contradiction.

Next, let us consider a clause with five literals $(a \vee b \vee c \vee d \vee e)$. Denoting $d \vee e$ by a new variable DE and using the breaking rule for four literals twice, we get

$$(a \vee b \vee c \vee d \vee e) = (a \vee b \vee c \vee DE)$$
$$= (a \vee b \vee x) \wedge (\neg x \vee c \vee DE)$$
$$= (a \vee b \vee x) \wedge (\neg x \vee c \vee y) \wedge (\neg y \vee d \vee e).$$

This leads to 3 new clauses with two new variables. We apply this transformation to each clause having more than 3 literals. Clearly this transformation takes polynomial time, since we traverse an original CNF-SAT and replace each clause with new clauses. We also need to make sure that we won't get an exponential number of new variables and clauses. During this transformation, a clause with m literals will be replaced by $(m - 2)$ clauses with $(m - 3)$ new variables. The number of clauses and variables is polynomially bounded. Thus, we have proved that the resulting 3-CNF formula is satisfiable if and only if CNF-SAT is satisfiable. ∎

9.3.2   Independent Set Problem

An independent set in an undirected graph G is a subset S of the vertices such that no pair of vertices in S is adjacent in G. The independent set problem (IS, in short) is to decide, for a given undirected graph G and natural number $k > 0$, whether G has an independent set of size k.

Theorem 3. The independent set problem *is $\mathcal{NP}$-complete.*

Proof. First, we show that IS $\in \mathcal{NP}$. Assume we have an independent set S. For each vertex in S we check every edge incident to it. If there is an edge that connects two vertices in S, the solution is not an independent set. Otherwise we accept S as the independent set.

In order to show that IS∈$\mathcal{NP}$-hard, we use a polynomial reduction from 3-SAT to IS. We will construct an undirected graph G from the 3-SAT instance with k clauses. The construction is based on the following ideas:

1. For each Boolean variable in 3-SAT we create a vertex in G.
2. All vertices corresponding to a given clause are connected to each other. This is because we want to make sure that only one vertex per clause is chosen in an independent set. This step creates k triangle subgraphs. If there are fewer than three literals, we can set the missing literals to any present literals.
3. Connect a vertex corresponding to a literal x to all vertices in G corresponding to its negation $\neg x$. We do not want to have two complementary vertices in one independent set.

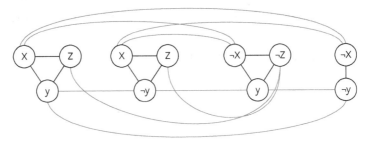

FIGURE 9.3   The graph constructed from $(x \vee y \vee z) \wedge (x \vee \neg y \vee z) \wedge (\neg x \vee y \vee \neg z) \wedge (\neg x \vee \neg y)$.

For example, figure 9.3 shows the graph constructed from 3-SAT formula $(x \vee y \vee z)$ $\wedge (x \vee \neg y \vee z) \wedge (\neg x \vee y \vee \neg z) \wedge (\neg x \vee \neg y)$. We conclude the construction with an observation that its runtime complexity is $O(k^2)$.

Claim: *3-SAT instance with k clauses is satisfiable if and only if the constructed graph G has an independent set of size k.*

Proof. We must show the implication in both directions.

⇒) Assume we have a truth assignment. Since the assignment makes each clause true, then at least one literal of each clause must be true. For some clauses we may have a few of such literals; we then arbitrarily pick one. Construct a set S of k vertices in G by choosing the vertex corresponding to the selected literal from each clause. S is an independent set. For the example in figure 9.4, let us choose the following truth assignment: $x = $ T, $y = $ F, and $z = $ F. The corresponded independent set S is shown in figure 9.5.

⇐) Suppose G has an independent set S of size k. Then S must include exactly one vertex from each clause. Also, S cannot have vertices representing a literal and its negation. We set all vertices in S to true. For vertices not in S, we choose the assignment arbitrarily. Thus, the independent set S yields a satisfying truth assignment.

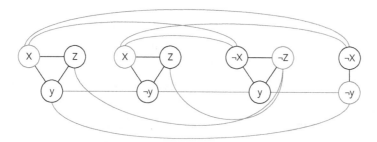

FIGURE 9.4   The independent set (shown in red) built from x = T, y = F and z = F.

Consider the independent set in figure 9.5. We set all green vertices to true (i.e., $z =$ T and $\neg x =$ T). The vertex y can be set either to true or false. It is easy to see that we have a truth assignment for $(x \vee y \vee z) \wedge (x \vee \neg y \vee z) \wedge (\neg x \vee y \vee \neg z) \wedge (\neg x \vee \neg y)$. ∎

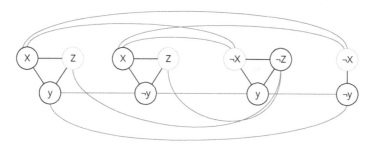

FIGURE 9.5   The independent set in green.

9.3.3   Vertex Cover Problem

The vertex cover problem (VC, in short) is to decide, for a given undirected graph G and natural number $k > 0$, whether G has a vertex cover of size k. In section 9.2 we showed that the independent set (IS) problem is polynomial time reducible to the vertex cover (VC) problem (i.e., IS $\leq_p$ VC). Therefore, combining this with the fact that IS $\in$ $\mathcal{NP}$-complete, and VC $\in \mathcal{NP}$, we conclude that VC is $\mathcal{NP}$-complete.

Let us restrict the instances of VC to undirected graphs with only even degree vertices. We will call this problem vertex cover even (VCE, in short).

Theorem 4. *The vertex* cover *even problem is $\mathcal{NP}$-complete.*

Proof. VCE $\in \mathcal{NP}$ follows immediately from the fact that VC $\in \mathcal{NP}$. VCE is the same problem as VC, only with more restrictions placed on the input.

In order to show that VCE is $\mathcal{NP}$-hard, we will use reduction from VC (i.e., VC $\leq$ p VCE. We need to convert any graph G into a graph with all even degree vertices G'. Note a simple fact that any undirected graph has an even number of odd degree vertices. Therefore, we construct G' by adding an extra vertex to G and connecting it to all vertices of odd degrees. See figure 9.6 for an example. ■

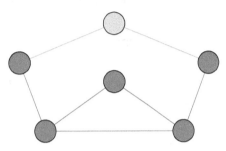

FIGURE 9.6   Graph G is in blue. We construct G' by adding a yellow vertex.

Claim: *G has a vertex cover of size k if and only if G' has a vertex cover of size $k + 1$.*

Proof. $\Rightarrow$) Assume G has a vertex cover of size k. Then the vertex cover of G' is created by adding the new vertex. Thus, the vertex cover size is $k + 1$.

$\Leftarrow$) Assume G' has a vertex cover of size $k + 1$. In order to get the vertex cover of G, we have to remove one vertex. Unfortunately, it does not always work. Consider G' in figure 9.6. The vertex cover of G' does not necessarily contain a yellow vertex. Assume that the vertex cover of G' is comprised of four blue vertices. If we remove one, we get a vertex cover of the smaller size but for a different graph—a graph with the yellow vertex. Its vertex cover is not identical to the vertex cover of G.

And so, the reduction we have described is not correct.

We revise our construction and add three new vertices to G. One of those new vertices is connected to all vertices of odd degrees. See figure 9.7.

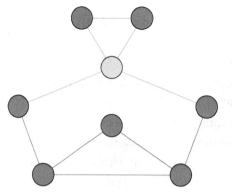

FIGURE 9.7   Graph G is in blue. We construct G' by adding three new vertices.

Claim: G has a vertex cover of size k if and only if G' has a vertex cover of size $k + 2$.

Proof. $\Rightarrow$) Assume $\bar{G}$ has a vertex cover of size k. Then the vertex cover of G' is created by adding two extra vertices (yellow and red in figure 9.5). Thus, the vertex cover size is $k + 2$.

$\Leftarrow$) Assume G' has a vertex cover of size $k + 2$. In order to get the vertex cover of G, we have to remove two vertices. Those two vertices are easily identified; they must from the set of extra vertices. In figure 9.7, we remove yellow and red vertices to get the right vertex cover for G. ∎

9.3.4   Graph Coloring Problem

The graph coloring problem is to decide, for a given undirected graph G and integer number $k > 0$, whether all vertices in G can be colored with k colors so that any two adjacent vertices are colored differently. There are certain classes of graphs when the coloring problem can be solved in polynomial time. One special case is when a graph is planar (see chapter 1.3.4). The second special case is $k = 2$, in which we are to decide if a graph is bipartite (see chapter 1.3.5). Unfortunately, for general graphs with $k \geq 3$, the problem is $\mathcal{NP}$-complete. To prove this, let us restrict the instances of the graph coloring problem to $k = 3$. We will call this problem 3-COLORING.

Theorem 4. *3-COLORING is $\mathcal{NP}$-complete.*

Proof. First, we show that 3-COLORING $\in \mathcal{NP}$. Assume we have a 3-color assignment S. For each vertex u in S we deterministically check its adjacent vertices. If there is an adjacent vertex of the same color as u, we reject this solution. Another way to prove that 3-COLORING $\in \mathcal{NP}$ is to non-deterministically guess an assignment of colors and then check each vertex. In either case this can be done in nondeterministic polynomial time.

In order to show that 3-COLORING $\in \mathcal{NP}$-hard, we use a polynomial reduction from 3-SAT. We will construct an undirected graph G from the 3-SAT instance. The construction is based on using "gadgets." A part of the original 3-SAT instance is translated into a "gadget" (a colored subgraph) that handles some details of the problem. These gadget subgraphs are then connected together to create a graph G. Our reduction consists of three gadgets. We associate the green-colored vertex with true and the red-colored vertex with false. We do not assign any special meaning to a blue vertex. The truth gadget is a triangle subgraph where each vertex has a different color. There will be only one truth gadget in G. The variable gadget is a subgraph with two vertices, colored either green or red. One vertex is associated with a variable, another with its negation. There will as many such gadgets as there are variables in the given 3-SAT instance. The variable gadgets are connected with the truth gadget via the blue vertex; see figure 9.8 for an example. Any 3-coloring of the that subgraph defines a valid truth assignment! And vice versa.

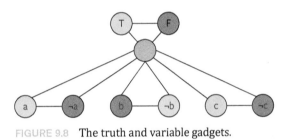

FIGURE 9.8   The truth and variable gadgets.

Finally, we have to make sure that the truth assignments satisfy the given clauses. This requires a new gadget for each clause. This gadget contains five unlabeled vertices that are connected with the truth and variable gadgets, as it's shown in figure 9.9.

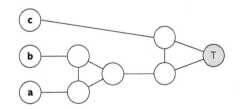

FIGURE 9.9   The clause gadget for $(a \vee b \vee c)$.

This gadget can be always colored with 3 colors except the case when all three literals, *a*, *b*, and *c*, are colored red (false). The proof is left as an exercise to the reader. Thus, if all the variables in a clause are false, the gadget cannot be 3-colored. On the other hand, if the clause gadget can be colored with 3 colors, then the associated clause in 3-SAT is satisfied.

Next, we put these gadgets together: Connect a truth gadget with the variable gadgets, connect the variable gadgets with the clause gadgets, and connect the clause gadgets with the truth gadget. As an example, the formula $(a \vee b \vee c)$ would be transformed into the graph shown in figure 9.10.

We conclude the construction noting that runtime complexity of building a graph G is $O(n)$, where n is the number of clauses in the original 3-SAT. The total number of vertices in G is also $O(n)$.

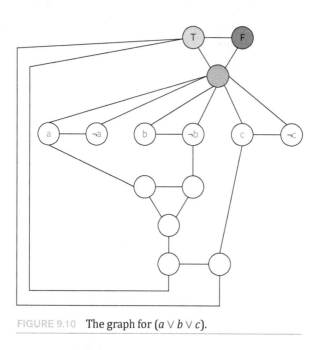

FIGURE 9.10   The graph for $(a \vee b \vee c)$.

Claim: *3-SAT instance is satisfiable if and only if G is 3-colorable.*

Proof. ⇒) Assume we have a truth assignment. In the constructed graph *G*, we color the variables with true or false according to the assignment. Coloring for the rest of vertices in the clause gadgets is forced.

⇐) Assume G is 3-colorable. We can extract a satisfying assignment from any 3-coloring by traversing the variable gadgets. We set a Boolean variable to true if in G it's colored green. We set a Boolean variable to false if in G it's colored red. ∎

REVIEW QUESTIONS

1. What is an algorithm?
2. What is the Church–Turing thesis?
3. What is a decision problem?
4. What is an undecidable problem?
5. What is the Halting problem?
6. What is the $\mathcal{P}$ versus $\mathcal{NP}$ conjecture?
7. (T/F) If someone proves $\mathcal{P} = \mathcal{NP}$, then it would imply that every decision problem can be solved in polynomial time.
8. (T/F) Any problem in $\mathcal{P}$ is also in $\mathcal{NP}$.
9. (T/F) Every decision problem is in $\mathcal{NP}$.
10. (T/F) Every problem in $\mathcal{NP}$ can be solved in exponential time by a deterministic Turing machine.
11. (T/F) All $\mathcal{NP}$-hard problems are in $\mathcal{NP}$.
12. (T/F) If a problem X can be reduced to linear programming in polynomial time, then X is in $\mathcal{P}$.
13. (T/F) If SAT $\leq_p$ A, then A is $\mathcal{NP}$-hard.
14. (T/F) If 3-SAT $\leq_p$ 2-SAT, then $\mathcal{P} = \mathcal{NP}$.
15. (T/F) If a problem $Y \leq_p X$, then it follows that $X \leq_p Y$.
16. (T/F) If A $\leq_p$ B and B is in $\mathcal{NP}$, then A is in $\mathcal{NP}$.
17. (T/F) If a problem X can be reduced to a known $\mathcal{NP}$-hard problem, then X must be $\mathcal{NP}$-hard.

EXERCISES

1. Prove that any optimization problem can be converted into a decision problem and vice versa.
2. Prove that if A $\leq_p$ B and B $\in \mathcal{NP}$ then A $\in \mathcal{NP}$.

3. Prove that if A $\leq_p$ B and B $\leq_p$ C then A $\leq_p$ C.

4. Prove that if Z $\leq_p$ Y and Y $\leq_p$ X then Z $\leq_p$ X.

5. Prove that the Halting problem is in $\mathcal{NP}$-hard class.

6. Assume that you are given a polynomial time algorithm that given a 3-SAT instance decides in polynomial time if it has a satisfying assignment. Describe a polynomial time algorithm that finds a satisfying assignment (if it exists) to a given 3-SAT instance.

7. Assume that you are given a polynomial time algorithm that decides if a directed graph contains a Hamiltonian cycle. Describe a polynomial time algorithm that outputs a sequence of vertices (in order) that form a Hamiltonian cycle.

8. Prove by reduction from 3-SAT that an integer linear program is $\mathcal{NP}$-complete.

9. The vertex cover problem (VC, in short) is to decide, for a given undirected graph G and natural number $k > 0$, whether G has a vertex cover of size k. Prove that VC is in $\mathcal{NP}$-complete class by reduction from 3-SAT. No other reductions can be used.

10. You are given a set S of n people and a set L of pairs of people that are mutually friends. Can these n people be seated for dinner around a circular table such that each person will sit next to friends on both sides? Prove that the problem (in short, DINNER) of finding such a sitting arrangement is $\mathcal{NP}$-complete.

11. Consider the 5-COLOR problem of deciding whether all vertices in undirected graph G can be colored with 5 colors so that any two adjacent vertices are colored differently. Prove that 5-COLOR is $\mathcal{NP}$-complete by reducing from 3-COLOR.

12. You are given an undirected connected graph $G = (V, E)$ in which a certain number of tokens $t(v) \geq 1$ placed on each vertex v. You will now play the following game. You pick a vertex u that contains at least two tokens, remove two tokens from u, and add one token to any one of adjacent vertices. The objective of the game is to perform a sequence of moves such that you are left with exactly one token in the whole graph. You are not allowed to pick a vertex with a 0 or 1 token. Prove that the problem of finding such a sequence of moves is $\mathcal{NP}$-complete by reduction from the Hamiltonian path.

13. We want to become celebrity chefs by creating a new dish. There are n ingredients and we'd like to use as many of them as possible. However, some ingredients don't go so well with others. There is $n \times n$ matrix D giving discord between any two ingredients, where $D[i,j]$ is a real value between 0 and 1. Any dish prepared

with these ingredients incurs a penalty, which is the sum of the discords between all pairs of ingredients in the dish. We would like the total penalty to be as small as possible. Consider the decision problem EXPERIMENTAL CUISINE: can we prepare a dish with at least k ingredients and with the total penalty at most p? Show that EXPERIMENTAL CUISINE is $\mathcal{NP}$-complete by giving a reduction from INDEPENDENT SET.

14. Given an undirected graph with positive edge weights, the BIG-HAM-CYCLE problem is to decide if it contains a Hamiltonian cycle C such that the sum of weights of edges in C is at least half of the total sum of weights of edges in the graph. Show that BIG-HAM-CYCLE is $\mathcal{NP}$-complete by reduction from the Hamiltonian cycle.

15. We know that finding a Hamiltonian cycle in a graph is $\mathcal{NP}$-complete. Show that finding a Hamiltonian path—a path that visits each vertex exactly once and isn't required to return to its starting point—is also $\mathcal{NP}$-complete.

16. Given a graph $G = (V, E)$ and a positive integer k, the longest-cycle problem is the problem of determining whether a simple cycle (no repeated vertices) of length k exists in a graph. Show that this problem is $\mathcal{NP}$-complete by reduction from the Hamiltonian cycle.

17. Given a graph $G = (V, E)$ and a positive integer k, the longest-path problem is the problem of determining whether a simple path (no repeated vertices) of length k exists in a graph. Show that this problem is $\mathcal{NP}$-complete by reduction from the Hamiltonian path.

18. You are given an undirected weighted graph $G = (V, E)$ with positive edge costs, a subset of vertices $R \subseteq V$, and a number C. Is there a tree in G that spans all vertices in R (and possibly some other in V) with a total edge cost of at most C? Prove that this problem is $\mathcal{NP}$-complete by reduction from vertex cover.

CPSIA information can be obtained
at www.ICGtesting.com
Printed in the USA
FSHW021335010421
80042FS